SpringerBriefs in Computer Science

Series Editors

Stan Zdonik
Peng Ning
Shashi Shekhar
Jonathan Katz
Xindong Wu
Lakhmi C. Jain
David Padua
Xuemin Shen
Borko Furht
V. S. Subrahmanian

For further volumes:
http://www.springer.com/series/10028

Alexandre Rademaker

A Proof Theory for
Description Logics

 Springer

Alexandre Rademaker
Applied Mathematics School
Fundação Getúlio Vargas
Praia de Botafogo 190
Rio de Janeiro
RJ 22250-900
Brazil

ISSN 2191-5768   ISSN 2191-5776 (electronic)
ISBN 978-1-4471-4001-6   ISBN 978-1-4471-4002-3 (eBook)
DOI 10.1007/978-1-4471-4002-3
Springer London Heidelberg New York Dordrecht

Library of Congress Control Number: 2012937297

Printed on acid-free paper

Springer is part of Springer Science+Business Media (www.springer.com)

To my family, Carla, Gabriela and Sofia. To my parents, André and Silvia whom I thank for their constant encouragement. I am specially thankful to my wife, Carla, for her love and friendship.

Preface

This book is the reprint of my thesis defended in 2010 at PUC-Rio. The content is mostly identical and contains only few adjustments and complements to the original thesis.

For instance, I took advance of some experiences gathered from the Semantic Web community during the 10th Semantic Web Conference at Bonn. A presentation during that event reminded me that I should write about how our systems can be used in the problem called "justification" by the Semantic Web community. I also add some brief comments about the extension of our deduction systems to deal with intuitionistic version of $\mathcal{ALC}$. This research was started right after I got my PhD and it is now referenced in the conclusion.

I specially thank Edward Hermann Haeusler, my advisor and friend, who never refuses to give me some advice and share his ideas. I also thank Leonardo Magalhes de Araujo Monteiro for his careful revision of the English language.

Rio de Janeiro   Alexandre Rademaker
January

Contents

Chapter 1
Introduction

Abstract Description Logics (**DLs**) are quite well-established as underlying logics for Knowledge Representation (**KR**). In a broader sense, a Knowledge Base (**KB**) specified in any description logic is called an Ontology. In this book, we are not interested in the technological concerns around Ontologies. For us, a DL theory, that is, a set of axioms in the **DL** logical language, and, an OWL file containing the same set of axioms is the same **KB**. Research in DL, since the beginning, was oriented to the development of systems and to their use in applications. Despite the higher efficiency of the recent available DL systems they do not provide to ontology engineers a good support for explanations of entailments like answering whether a subsumption holds or not, query or classification of concepts and roles of an ontology. This book is our first step in the direction of contract deduction systems with better explanation support.

Keywords Description logic · ALC · Deduction system · Proof theory · Proof explanation · Sequent calculus · Natural deduction

1.1 Description Logics

Description Logics (**DLs**) are quite well-established as underlying logics for Knowledge Representation (**KR**). Part of this success comes from the fact that it can be seen as one (logical) successor of Semantics Networks [28], Frames [25] and Conceptual Graphs [38] and as well as, an elegant and powerful restriction of **FOL** by guarded prefixes, that also leads to a straight interpretation into the **K** propositional modal logic.

The core of the **DLs** is the $\mathcal{ALC}$ description logic. In a broader sense, a Knowledge Base (**KB**) specified in any description logic having $\mathcal{ALC}$ as core is called an Ontology. In this book we will not take any ontological[1] discussion on the choice for this terminology by the computer science community. Moreover, we are not inter-

---

[1] In the philosophical sense.

A. Rademaker, *A Proof Theory for Description Logics*,
SpringerBriefs in Computer Science, DOI: 10.1007/978-1-4471-4002-3_1,
© The Author(s) 2012

ested in the technological concerns around Ontologies, the **Web** or the fact that there is an XML dialect for writing Ontologies, just named OWL [14]. For us, a DL theory presentation, that is, a set of axioms in the **DL** logical language, and, an OWL file containing the same set of axioms is the same **KB**.

Description Logics is a family of formalisms used to represent knowledge of a domain. In contrast with other knowledge representation systems, Description Logics are equipped with a formal, logic-based semantics. This logic-based semantics provides to systems based on it various inference capabilities to deduce implicit knowledge from the explicitly represented knowledge.

1.2 Motivation

Research in DL, since the beginning, was oriented to the development of systems and to their use in applications. In the first half of the 1980's several systems were developed including KL-ONE [6] and KRYPTON [5], only to mention two. They were called first generation DL systems. Later, in the second half of 1980's, the second generation of DL systems appears, the BACK [19], CLASSIC [4] and LOOM [22] systems.

In the last years several DL systems have been developed incorporating different DL fragments but similar with respect to the underlying reasoning algorithm. Nowadays, DL has good reasoners from the point of view of providing yes/no answers or various inference tasks like subsumption of concepts (see Chap. 2) or classification.[2] We mention the open-source Pellet [37], Racer Pro [15] and Fact [39].[3]

The first DL systems implement structural subsumption algorithms [36]. The basic idea underlying structural subsumption is to transform terms into *canonical normal forms*, which are then structurally compared. Structural subsumption algorithms are therefore also referred to as normalize-compare algorithms. There is one important drawback of normalize-compare algorithms. That is, in general it is straightforward to prove the correctness of such algorithms but there is no method for proving their completeness [27].

As far as we know, the most well-known existing DL reasoners implement variations of Tableaux proof procedure for DL [8, 9, 35]. As pointed in [27], Tableaux procedures for computing subsumption of concepts had the advantage of providing good basis for theoretical investigations. Not only was their correctness and completeness easy to prove, they also allowed a systematic study of the decidability and the tractability of different DL dialects. On the other hand, the main disadvantage of tableaux-based algorithms is that they are not constructive but rather employ refutation techniques. That is, in order to prove $\alpha \sqsubseteq \beta$, it is proved that the concept $\alpha \sqcap \neg \beta$ is not satisfiable (see Chap. 2).

---

[2] The classification checks subsumption between the terms defined in the terminology and computes the subsumption hierarchy of them.

[3] A possible outdated list is maintained in the Description Logics website http://dl.kr.org/.

As claimed by [23], the use of Description Logics by regular users, that is, non-technical users, would be wider if the computed inferences could be presented as a natural language text—or any other presentation format at the domain's specification level of abstraction—without requiring any knowledge on logic to be understandable.

Despite the higher efficiency of the recent available DL systems they do not provide to ontology engineers a good support for explanations on their two main uses, namely, answering whether a subsumption holds or not, and, a classification result.

Some works [21, 23, 24] describe methods to extract explanations from DL-Tableaux proofs. Particularly, [7] describes the explanation extraction in quite few details, making impossible a feasible comparison with [21, 24]. In [3], for example, it is described a Sequent Calculus (SC) obtained by a standard transformation from Tableaux into SC systems applied to the DL-tableaux described in [35]. Hustadt and Schmidt [20] presents also a Resolution procedure for DL but does not address explanation extraction. It is worth noticing that the DL-tableaux do not implement non-analytic cuts, and hence proof resulted from this transformation is a cut-free proof. Moreover, even when dealing with the **TBOX** (see Chap. 2) the SC just discussed strongly deals with individuals, the **ABOX** aspect of an Ontology.

Simple Tableaux procedures are those not able to implement non-analytic cuts. The Tableaux procedures used for $\mathcal{ALC}$ are simple. It is also known that Simple Tableaux cannot produce always short proofs,[4] that is, polynomially lengthy proofs, concerning the combined length[5] of its conclusion and set of (used) axioms from the Ontology. This is an easy corollary of the theorem that asserts that Simple Tableaux as well Resolution cannot produce short proofs for the Pigeonhole Principle (PHP) [18]. PHP is easily expressed in propositional logic, and hence, is also easily expressed in $\mathcal{ALC}$. On the other hand, Sequent Calculus (SC) (with the cut rule) has short proofs for PHP. In [10, 13] it is shown, distinct, SC proof procedures that incorporate mechanisms that are somehow equivalent to the cut-rule. Anyway, both articles show how to obtain short proof, in SC, for the PHP. We believe that super-polynomial proofs, like the ones generated by simple Tableaux, cannot be considered as good sources for text generation. The reader might want to consider that only the reading of the proof itself is a super-polynomial task regarding time complexity.

The final consideration worth of mentioning regarding a motivation to obtain a Natural Deduction system for $\mathcal{ALC}$, despite providing a variation of themes, is the possibility of getting ride on a weak form of the Curry-Howard isomorphism in order to provide explanations with greater content. This last affirmative takes into account that the reading (explanatory) content of a proof is a direct consequence of its computational content. This is discussed in Chaps. 5 and 7.

A last observation lies on the fact that allowing this incremental proof-theoretical design of systems to **DL** we obtain a uniform specialization of the general proof procedure for $\mathrm{ND}_{\mathcal{ALC}}$.

---

[4] If we consider the assumption that **NP** $\neq$ **CO-NP**.

[5] Number of symbols.

1.3 What This Book is About

In this book, we present two deduction systems for $\mathcal{ALC}$[6] and $\mathcal{ALCQI}$,[7] a sequent calculus and a natural deduction system. The first motivation for developing such systems is the extraction of computational content of $\mathcal{ALC}$ and $\mathcal{ALCQI}$ proofs. More precisely, these systems were developed to allow the use of natural language to render a Natural Deduction proof. The sequent calculi were intermediate steps towards a Natural Deduction Systems [26].

Our main motivation to develop such systems are that natural language rendering of a Natural Deduction proofs is worthwhile in some contexts like, for instance, proof of conformance in security standards [1].

Our Sequent Calculus system is also compared with other approaches like Tableaux [35] and the Sequent Calculus for $\mathcal{ALC}$ [2, 11, 23] based on this very Tableaux. In fact, our system does not use individual variables (first-order ones) at all. The main mechanism in our system is based on labeled formulas. The labeling of formulas is among one of the most successful artifacts for keeping control of the context in the many existent quantification in logical system and modalities. For a detailed reading on this approach, we point out [12, 17, 32–34].

Our Sequent Calculi systems argue in favor of better explanation schemata obtained from proofs, regarding those obtained from a $\mathcal{ALC}$-Tableaux. Both systems do not use individuals, producing a purely conceptual reasoning for TBOX. Moreover, it is worth mentioning that both systems can also provide proofs with cuts, as opposed to the one presented in [23].

1.4 How This Book is Organized

Chapter 2 presents some background introducing DL languages and semantics.

Chapter 3 presents the system $SC_{\mathcal{ALC}}$, a sequent calculus for $\mathcal{ALC}$ and proves that it is sound and complete. This chapter was originally published in [29, 31], where we proved that $SC_{\mathcal{ALC}}$ has the desirable property of allowing the construction of cut-free proofs. That is, we prove that the *cut rule* can be eliminated from the system $SC_{\mathcal{ALC}}$ without lost the completeness and soundness.

In Chap. 4, we compare $SC_{\mathcal{ALC}}$ with the Structural Subsumption algorithm and the Tableaux for $\mathcal{ALC}$. The comparison is made regarding: (1) the proof construction procedure in Structural Subsumption algorithm and $SC_{\mathcal{ALC}}$; and (2) the ability of $\mathcal{ALC}$-Tableaux to construct counter-models. The results from this chapter were first published in [30].

---

[6] $\mathcal{ALC}$ means Attributive Language with Complements, a basic Description Logic.

[7] The $\mathcal{Q}$ in $\mathcal{ALCQ}$ means the introduction in the language of qualified number restriction constructors.

In Chap. 5 we present the Natural Deduction for $\mathcal{ALC}$ named $\mathrm{ND}_{\mathcal{ALC}}$. In this chapter, we also prove that $\mathrm{ND}_{\mathcal{ALC}}$ is sound and complete. We also proof the normalization theorem for $\mathrm{ND}_{\mathcal{ALC}}$. The results in this chapter were published in [16].

In Chap. 6, we present the extensions of our Natural Deduction and Sequent Calculus for $\mathcal{ALC}$ to $\mathcal{ALCQI}$. We prove the soundness of both systems and some ongoing work regarding their completeness. In Chap. 7, we present the motivation and some discussion about the extraction of explanations from proofs. We compare proofs in Tableaux, Natural Deduction and Sequent Calculi. Also in this chapter, we present our Natural Deduction for $\mathcal{ALCQI}$ to reasoning over a UML diagram. The example helps us compare how proofs in $\mathrm{ND}_{\mathcal{ALCQI}}$ can be easier explained than proofs using Tableaux.

In Chap. 8, we present a prototype theorem prover that implements our Natural Deduction and Sequent Calculi systems. Finally, in Chap. 9, we present some conclusions and further work.

References

1. do Amaral, F.N., Bazílio, C., da Silva, G.M.H., Rademaker, A., Haeusler, E.H.: An ontology-based approach to the formalization of information security policies. EDOCW **0**, 1 (2006)
2. Borgida, A., Franconi, E., Horrocks, I., McGuinness, D., Patel-Schneider, P.: Explaining $\mathcal{ALC}$ subsumption. In: Proceedings of the International Workshop on Description Logics, pp. 33–36 (1999)
3. Borgida, A., Franconi, E., Horrocks, I., McGuinness, D.L., Patel-Schneider, P.F.: Explaining ALC subsumption. In: Lambrix, P., Borgida, A., Lenzerini, M., Möller, R., Patel-Schneider, P.F. (eds.) Proceedings of the 1999 International Workshop on Description Logics 1999, vol. 22. Linköping, Sweden (1999). http://SunSITE.Informatik.RWTH-Aachen.DE/Publications/CEUR-WS/Vol-22/borgida.ps
4. Brachman, R., McGuiness, D.L., Patel-Schneider, P.F., Resnick, L.A., Borgida, A.: Living with classic: when and how to use a kl-one-like language. In: Sowa, J. (ed.) Principles of Semantic Networks: Explorations in the Representation of Knowledge, pp. 401–456. Morgan Kaufmann, San Mateo (1991)
5. Brachman, R.J., Fikes, R.E., Levesque, H.J.: Krypton: A functional approach to knowledge representation. IEEE Comput. **16**, 67–73 (1983)
6. Brachman, R.J., Schmolze, J.: An overview of the kl-one knowledge representation system. Cogn. Sci. **9**(2) (1985)
7. Deng, X., Haarslev, V., Shiri, N.: Using patterns to explain inferences in $\mathcal{ALCHI}$. Computat. Intell. **23**(3), 386–406 (2007)
8. Donini, F., Lenzerini, M., Nardi, D., Nutt, W.: The complexity of concept languages. Inform. Computat. **134**(1), 1–58 (1997)
9. Donini, F.M., Lenzerini, M., Nardi, D., Nutt, W.: The complexity of concept languages. Inform. Computat. **134**(1), 1–58 (1991)
10. Finger, M., Gabbay, D.: Equal rights for the cut: computable non-analytic cuts in cut-based proofs. Log. J. IGPL **15**(5–6), 553–575 (2007)
11. Fitting, M.: Proof Methods for Modal and Intuitionistic Logics. Reidel, Dordrecht (1983)
12. Gabbay, D.M.: Labelled Deductive Systems, vol. 1. Oxford University Press, Oxford (1996)
13. Gordeev, L., Haeusler, E., Costa, V.: Proof compressions with circuit-structured substitutions. In: Zapiski Nauchnyh Seminarov POMI (2008), to appear
14. Group, W.O.W.: Owl 2 web ontology language document overview (2009). http://www.w3.org/TR/owl2-overview/

15. Haarslev, V., Möller, R.: Racer system description. In: Goré, R., Leitsch, A., Nipkow, T. (eds.) International Joint Conference on Automated Reasoning, IJCAR'2001, June 18–23, Siena, Italy, pp. 701–705. Springer, Berlin (2001)

16. Haeusler, E.H., Rademaker, A.: Is it important to explain a theorem? a case study on uml and $\mathcal{ALCQI}$. In: ER '09: Proceedings of the ER 2009 Workshops (CoMoL, ETheCoM, FP-UML, MOST-ONISW, QoIS, RIGiM, SeCoGIS) on Advances in Conceptual Modeling–Challenging Perspectives, pp. 34–44. Springer, Berlin, (2009). DOI: 10.1007/978-3-642-04947-7_6

17. Haeusler, E.H., Renteria, C.J.: A natural deduction system for CTL. Bull. Sect. Log. **31**(4), 231 (2002)

18. Haken, A.: The intractability of resolution. Theoret. Comput. Sci. **39**, 297–308 (1985)

19. Hoppe, T., Kindermann, C., Quantz, J.J., Schmiedel, A., Fischer, M.: Back v5 tutorial & manual. Technical Report, KIT Report 100, Technische Universitat, Berlin (1993)

20. Hustadt, U., Schmidt, R.A.: Issues of decidability for description logics in the framework of resolution. In: Automated Deduction in Classical and Non-Classical Logics, pp. 191–205. Springer, Berlin (2000)

21. Liebig, T., Halfmann, M.: Explaining subsumption in $\mathcal{ALEHF}_{R+}$ tboxes. In: Horrocks, I., Sattler, U., Wolter, F. (eds.) Proceedings of the 2005 International Workshop on Description Logics–DL2005, pp. 144–151. Edinburgh, Scotland (2005)

22. MacGregor, R.: Using a description classifier to enhance deductive inference. In: Proceedings of Seventh IEEE Conference on AI Applications, pp. 141–147. Miami, Florida (1991)

23. McGuinness, D.L.: Explaining reasoning in description logics. Ph.D. Thesis, Rutgers University, New Brunswick (1996)

24. McGuinness, D.L., Borgida, A.: Explaining subsumption in description logics. In: International Joint Conference on Artificial Intelligence, vol. 14, pp. 816–821 (1995)

25. Minsky, M.: A framework for representing knowledge. In: Winston, P.H. (ed.) The Psychology of Computer Vision, pp. 211–277. McGraw-Hill, New York (1975)

26. de Oliveira, D.A.S., de Souza, C.S., Haeusler, E.H.: Structured argument generation in a logic based kb-system. In: Moss, L.S., Ginzburg, J., de Rijke, M. (eds.) Logic Language and Computation, no. 96 in CSLI Lecture Notes, 1 edn., pp. 237–265. CSLI, Stanford, California (1999)

27. Quantz, J.J., Dunker, G., Bergmann, F., Kellner, I.: The flex system. Technical Report, Kit Report 124, Technische Universitat, Berlin (1995)

28. Quillian, M.R.: Semantic memory. In: Minsky, M. (ed.) Semantic Information Processing, pp. 216–270. MIT Press, Cambridge (1968)

29. Rademaker, A., do Amaral, F.N., Haeusler, E.H.: A sequent calculus for $\mathcal{ALC}$. In: Monografias em Ciência da Computação 25/07, Departamento de Informática, PUC-Rio (2007)

30. Rademaker, A., Haeusler, E.H.: Toward short and structural $\mathcal{ALC}$-reasoning explanations: a sequent calculus approach. In: Proceedings of Brazilian Symposium on Artificial Intelligence. Advances in Artificial Intelligence–SBIA 2008, pp. 167–176. Springer, Berlin (2008). 10.1007/978-3-540-88190-2_22

31. Rademaker, A., Haeusler, E.H., Pereira, L.C.: On the proof theory of $\mathcal{ALC}$. In: The Many Sides of Logic. Proceedings of 15th Brazilian Logic Conference. College Publications, London (2008). A resumed version is available at http://www.cle.unicamp.br/e-prints/vol_8,n_6,2008. html

32. Renteria, C.J., Haeusler, E.H.: A natural deduction system for keisler logic. Electron. Notes Theoret. Comput. Sci. **123**, 229–240 (2005)

33. Renteria, C.: Uma abordagem geral para quantificadores em dedução natural. Ph.D. Thesis, PUC-Rio, DI (2000)

34. Renteria, C.J., Haeusler, E., Veloso, P.: NUL: Natural deduction for ultrafilter logic. Bull. Sect. Log. **32**(4), 191–200 (2003)

35. Schmidt-Schauß, M., Smolka, G.: Attributive concept descriptions with complements. Artif. Intell. **48**(1), 1–26 (1991)

36. Schmolze, J., Israel, D.: Kl-one: Semantics and classification. Technical Report 5421, BBN (1983)

37. Sirin, E., Parsia, B., Grau, B.C., Kalyanpur, A., Katz, Y.: Pellet: A practical owl-dl reasoner. J. Web Semant. **5**(2), 51–53 (2007)
38. Sowa, J.F. (ed.): Principles of Semantic Networks: Explorations in the Representation of Knowledge. Morgan Kaufmann, Los Altos (1991)
39. Tsarkov, D., Horrocks, I.: Fact++ description logic reasoner: System description. LNCS 4130 LNAI, 292–297 (2006)

Chapter 2
Background

Abstract Description Logics is a family of knowledge representation formalisms used to represent knowledge of a domain, usually called "world". For that, it first defines the relevant concepts of the domain—"terminology"—and then, using these concepts, specifies properties of objects and individuals of that domain. In this chapter, we review the syntax, semantics and the main logical properties of DL that we will use in the following chapters.

Keywords Description logic · ALC · Logic · Syntax · Semantics · Proof theory · Axiomatization · Theory · Satisfiable · Tautology

2.1 A Basic Description Logic

Comparing to its predecessors formalisms, Description Logics are equipped with a formal, logic-based semantics. Description Logics differ each other from the constructors they provide. Concept constructors are used to build more complex descriptions of concepts from *atomic concepts* and role constructor to build complex role descriptions from *atomic roles*.

$\mathcal{ALC}$ is a basic Description Logics [1] and its syntax of concept descriptions is as following:

$$\phi_c \rightarrow \top \mid \bot \mid A \mid \neg\phi_c \mid \phi_c \sqcap \phi_c \mid \phi_c \sqcup \phi_c \mid \exists R.\phi_c \mid \forall R.\phi_c$$

where A stands for atomic concepts and R for atomic roles. The concepts $\bot$ and $\top$ could be omitted since they are just abbreviations for $\alpha \sqcap \neg\alpha$ and $\alpha \sqcup \neg\alpha$ for any given concept description α.

The semantics of concept descriptions is defined in terms of an *interpretation* $\mathcal{I} = (\Delta^{\mathcal{I}}, \cdot^{\mathcal{I}})$. The domain $\Delta^{\mathcal{I}}$ of $\cdot^{\mathcal{I}}$ is a non-empty set of individuals and the interpretation function $\cdot^{\mathcal{I}}$ maps each atomic concept A to a set $A^{\mathcal{I}} \subseteq \Delta^{\mathcal{I}}$ and for

A. Rademaker, *A Proof Theory for Description Logics*,
SpringerBriefs in Computer Science, DOI: 10.1007/978-1-4471-4002-3_2,
© The Author(s) 2012

each atomic role a binary relation $R^{\mathcal{I}} \subseteq \Delta^{\mathcal{I}} \times \Delta^{\mathcal{I}}$. The function $.^{\mathcal{I}}$ is extended to concept descriptions inductive as follows:

$$\top^{\mathcal{I}} = \Delta^{\mathcal{I}}$$
$$\bot^{\mathcal{I}} = \emptyset$$
$$(\neg C)^{\mathcal{I}} = \Delta^{\mathcal{I}} \setminus C^{\mathcal{I}}$$
$$(C \sqcap D)^{\mathcal{I}} = C^{\mathcal{I}} \cap D^{\mathcal{I}}$$
$$(C \sqcup D)^{\mathcal{I}} = C^{\mathcal{I}} \cup D^{\mathcal{I}}$$
$$(\exists R.C)^{\mathcal{I}} = \{a \in \Delta^{\mathcal{I}} \mid \exists b.(a, b) \in R^{\mathcal{I}} \wedge b \in C^{\mathcal{I}}\}$$
$$(\forall R.C)^{\mathcal{I}} = \{a \in \Delta^{\mathcal{I}} \mid \forall b.(a, b) \in R^{\mathcal{I}} \rightarrow b \in C^{\mathcal{I}}\}$$

Knowledge representation systems based on description logics provide various inference capabilities that deduce implicit knowledge from the explicitly represented knowledge. One of the most important inference services of DL systems is computing the subsumption hierarchy of a given finite set of concept descriptions.

Definition 1 The concept description D subsumes the concept description C, written $C \sqsubseteq D$, if and only if $C^{\mathcal{I}} \subseteq D^{\mathcal{I}}$ for all interpretations $\mathcal{I}$.

Definition 2 C is satisfiable if and only if there is an interpretation $\mathcal{I}$ such that $C^{\mathcal{I}} \neq \emptyset$.

Definition 3 C is valid or a tautology if and only if, for all interpretation $\mathcal{I}$, $C^{\mathcal{I}} \equiv \Delta^{\mathcal{I}}$.

Definition 4 C and D are equivalent, written $C \equiv D$, if and only if $C \sqsubseteq D$ and $D \sqsubseteq C$.

We used to call $C \sqsubseteq D$ and $C \equiv D$ *terminological axioms*. Axioms of the first kind are called *inclusions*, while axioms of the second kind are called *equalities*. If an interpretation satisfies an axiom (or a set of axioms), then we say that is a *model* of this axiom (or set of axioms).

An equality axiom whose left-hand side is an atomic concept is a *definition*. Definitions are used to introduce *names* for complex descriptions. For instance, the axiom

$$Mother \equiv Woman \sqcap \exists\, hasChild.Person$$

associates to the description on the right-hand side the name *Mother*.

A finite set of definitions $\mathcal{T}$ where no symbolic name is defined more than once is called a *terminology* or *TBox*. In other words, for every atomic concept A there is at most one axiom in $\mathcal{T}$ whose left-hand side is A. Given a $\mathcal{T}$, we divide the atomic concepts occurring in it into two sets, the *name symbols* $\mathcal{N}_{\mathcal{T}}$ that occur on the left-hand side of some axiom and the *base symbols* $\mathcal{B}_{\mathcal{T}}$ that occur only on the right-hand side of axioms. Name symbols are also called *defined* concepts and base symbols *primitive* concepts. The terminology should *define* the name symbols in terms of the base symbols.

With the definitions of the previous paragraphs, we must also extend the definitions of *interpretations* to deal with TBox. A *base interpretation* $\cdot^{\mathcal{I}}$ for $\mathcal{T}$ is an interpretation just for the base symbols. An interpretation that also interprets the name symbols is called an *extension* of $\cdot^{\mathcal{I}}$. There are much more to say about such extensions. For instance, whenever we have cyclic definitions in a *TBox* the *descriptive* semantics given so far is not sufficient. In that case, we usually work with *fixpoint semantics*, we cite [1] for a complete reference.

2.2 Individuals

Besides the TBox component, in a knowledge base we usually have to describe individuals and assertions about them. We call the set of assertions about individual in a knowledge base a *world description* or *ABox*. In a ABOX we introduce individuals and describe their properties using the roles and concepts introduced or defined in the TBox. We have two kind of formulas to express assertions about individuals:

$$C(a) \qquad R(b, c)$$

The formula on the left is called *concept assertion*. It states that the individual a belongs to the interpretation of the concept C. The formula on the right is called *role assertion* that states that the individual c is a filler of the role R for b. Following the typical example from [1], if *Father* is a concept name and *hasChild* a role name, then we can have the following assertions about individual named *Peter, Paul, Mary*:

$$Father(Peter) \qquad hasChild(Mary, Paul)$$

The meaning of the left assertion is that *Peter* is a father and the assertion on the right says that *Paul* is a child of *Mary*.

Once more we have to extend the notion of interpretation in order to provide semantics to ABoxes. Essentially, the interpretation $\mathcal{I} = (\Delta, \cdot^{\mathcal{I}})$ besides mapping concepts to sets and roles to binary relations, also maps individual names a to an element $a^{\mathcal{I}} \in \Delta^{\mathcal{I}}$. We usually assume that distinct names denote distinct objects, this is called the *unique name assumption* (UNA). Formally, if a and b are distinct names, then $a^{\mathcal{I}} \neq b^{\mathcal{I}}$.

An interpretation $\mathcal{I}$ satisfy the assertion $C(a)$, if $a^{\mathcal{I}} \in C^{\mathcal{I}}$ and the role assertion $R(a, b)$, if $(a^{\mathcal{I}}, b^{\mathcal{I}}) \in R^{\mathcal{I}}$. In that cases, we write:

$$\mathcal{I} \models C(a) \qquad \mathcal{I} \models R(a, b)$$

An interpretation satisfies an ABox if it satisfies each assertions on it, that is, it is a *model* for the ABox. An interpretation that satisfies an ABox with respect to a TBox whenever it is a model for both.

2.3 Description Logics Family

If we add to $\mathcal{ALC}$ more constructors, more expressivity power to describe concepts and roles we obtain. Description logics are a huge family of logics, it is not our goal to present and discuss all of them. We will describe in this section only the extensions of $\mathcal{ALC}$ that we will deal in this book. For a complete reference we indicate [1].[1]

Two of the most useful extensions of $\mathcal{ALC}$ is $\mathcal{ALCN}$ and $\mathcal{ALCQ}$. $\mathcal{ALCN}$ includes *number restrictions* written as $\leq nR$ or $\geq nR$ where n ranges over non-negative integers. $\mathcal{ALCQ}$ allows constructors for qualified number restrictions of the form $\leq nR.C$ and $\geq nR.C$. The semantics of those constructors are given by the definitions below.

$$(\leq nR)^{\mathcal{I}} = \{a \in \Delta^{\mathcal{I}} \mid |\{b \mid (a,b) \in R^{\mathcal{I}}\}| \leq n\}$$
$$(\geq nR)^{\mathcal{I}} = \{a \in \Delta^{\mathcal{I}} \mid |\{b \mid (a,b) \in R^{\mathcal{I}}\}| \geq n\}$$
$$(\leq nR.C)^{\mathcal{I}} = \{a \in \Delta^{\mathcal{I}} \mid |\{b \mid (a,b) \in R^{\mathcal{I}} \wedge b \in C^{\mathcal{I}}\}| \leq n\}$$
$$(\geq nR.C)^{\mathcal{I}} = \{a \in \Delta^{\mathcal{I}} \mid |\{b \mid (a,b) \in R^{\mathcal{I}} \wedge b \in C^{\mathcal{I}}\}| \geq n\}$$

We name $\mathcal{AL}$-languages using letters to indicate the allowed constructor:

$$\mathcal{AL}[\mathcal{U}][\mathcal{E}][\mathcal{N}][\mathcal{Q}][\mathcal{C}]$$

The $\mathcal{AL}$ language is a restriction of $\mathcal{ALC}$ without union of concept ($\sqcup$), negation is only allowed to atomic concepts and limited existential quantification, that is, existential quantification only over $\top$ concept ($\exists R.\top$). The $\mathcal{U}$ stands for union of concepts, $\mathcal{E}$ for full existential quantification, $\mathcal{N}$ for number restrictions, $\mathcal{Q}$ for qualified number restrictions and $\mathcal{C}$ for full negation of concepts (not only atomics ones).

Taking into account the semantics, some of these languages are equivalent. For example, the semantics forces the equivalent between $C \sqcup D \equiv \neg(\neg C \sqcap \neg D)$ and $\exists R.C \equiv \neg \forall R.\neg C$. That is, union and full existential quantification can be expressed using negation and vice versa. That is why we use $\mathcal{ALC}$ instead of $\mathcal{ALUE}$ and $\mathcal{ALCN}$ instead of $\mathcal{ALUEN}$. One can also observe that $\mathcal{ALCN}$ is superseded by $\mathcal{ALCQ}$. That is, if we limit the qualified number restrictions of $\mathcal{ALCQ}$ to the $\top$ concept allowing only $\leq nR.\top$ and $\geq nR.\top$, we obtain $\mathcal{ALCN}$.

As said before, description logics are a huge family of formalisms. Much more constructors were introduced in the basic $\mathcal{ALC}$ to express: role constructors; concrete domains; modal, epistemic and temporal operators; *fuzzy* and probabilities to express uncertain or vague knowledge to cite just some of them [1]. Nevertheless, the languages presented so far will be sufficient for us in this work.

---

[1] We also point to the Description logics website at http://dl.kr.org/.

2.4 Reasoning in DLs

A knowledge base—TBox and ABox—equipped with its semantics is equivalent to a set of axioms in first-order predicate logic. Thus, as said before, like any other set of axioms, it contains implicit knowledge that *logical inferences* can make explicit.

When we are constructing a TBox $\mathcal{T}$, by defining new concepts, possibly in terms of others that have been defined before, it is important to enforce the consistence of the TBox. That is, it is important that new concepts make sense and do not be contradictory with old ones. Formally, a concept makes sense if there is some interpretation that satisfies the axioms of $\mathcal{T}$ such that the concept denotes a nonempty set in that interpretation.

Definition 5 (*Satisfiability*) A concept C is *satisfiable* with respect to $\mathcal{T}$ if there is a model $\cdot^{\mathcal{I}}$ of $\mathcal{T}$ such that $C^{\mathcal{I}}$ is nonempty. In this case, $\cdot^{\mathcal{I}}$ is a *model* of C.

While modeling a domain of knowledge into a TBox other important inference service is necessary. For instance, it is usually interesting to organize the concepts of a TBox into a taxonomy. That is, it is important to know whether some concept is more general than another one: the *subsumption problem*. Furthermore, other interesting relationships among concepts is the *equivalence*.

Definition 6 (*Subsumption*) A concept C is *subsumed* by a concept D with respect to $\mathcal{T}$ if $C^{\mathcal{I}} \subseteq D^{\mathcal{I}}$ for every model $\cdot^{\mathcal{I}}$ of $\mathcal{T}$. In this case we write $C \sqsubseteq_{\mathcal{T}} D$ or $\mathcal{T} \models C \sqsubseteq D$.

Definition 7 (*Equivalence*) Two concepts C and D are *equivalent* with respect to $\mathcal{T}$ if $C^{\mathcal{I}} = D^{\mathcal{I}}$ for every model $\cdot^{\mathcal{I}}$ of $\mathcal{T}$. In this case we write $C \equiv_{\mathcal{T}} D$ or $\mathcal{T} \models C \equiv D$.

If the TBox is clear from the context or empty we can drop the qualification and simply write $\models C \sqsubseteq D$ if C is subsumed by D, and $\models C \equiv D$ if they are equivalent.

Since it is not our main concern in this book, we will not go into more details about the equivalence and reductions between reasoning problems in Description Logics. Basically, the different kinds of reasoning can be all reduced to a main inference problem named the consistency check for ABox [1].

2.5 Inference Algorithms

There are two main algorithms to reasoning in Description Logics: *structural subsumption algorithms* and *tableaux-based algorithms* [1]. We postpone the presentation of these two algorithms for Chap. 4, here we will just present briefly comments about them. One of the differences between them relies on the logical languages that each one can handle.

For the description logic $\mathcal{ALN}$ and its subsets, that is, the Description Logic not allowing full negation ($\neg C$), disjunction ($C \sqcup D$) nor full existential ($\exists R.C$), the subsumption of concepts can be computed by structural subsumption algorithms. The idea of these algorithms is compare the syntactic structure of concept descriptions. These algorithms are usually very efficient, polynomial time complexity [2] indeed.

For $\mathcal{ALC}$ and its extensions, the satisfiability of concepts and the subsumption of concept usually can be computed by *tableau-based algorithms* which are sound and complete for these problems [1]. The first tableau-based algorithm for satisfiability of $\mathcal{ALC}$-concepts was presented by [4]. As we said before, some reasoning problems in Description Logics can be reduced to others, in special, the problem to test the subsumption of concepts is reduced to the problem of test the (un)satisfiability of a concept description. These algorithms use the fact that $C \sqsubseteq D$ if and only if $C \sqcap \neg D$ is unsatisfiable [1]. Regarding the complexity, the tableau-based satisfiability algorithm for $\mathcal{ALC}$ is a PSPACE-hard problem [4].

2.6 $\mathcal{ALC}$ Axiomatization

From [3] we know that $\mathcal{ALC}$ is sound and complete for any Classical Propositional Logic axiomatization containing the axioms:

Definition 8 (*An Axiomatization of $\mathcal{ALC}$*)

$$\forall R.(\alpha \sqcap \beta) \equiv \forall R.\alpha \sqcap \forall R.\beta \qquad (2.1)$$

$$\forall R.\top \equiv \top \qquad (2.2)$$

As usual, $\exists R.\alpha$ can be taken as a shorthand for $\neg \forall R.\neg \alpha$, as well as $\forall R.\alpha$ as a shorthand for $\neg \exists R.\neg \alpha$. Taking $\exists R.\alpha$ as a definable concept, the axioms change to

$$\exists R.(\alpha \sqcup \beta) \equiv \exists R.\alpha \sqcup \exists R.\beta \qquad (2.3)$$

$$\exists R.\bot \equiv \bot \qquad (2.4)$$

The following rule, also known as necessitation rule:

$$\frac{\vdash \alpha}{\vdash \forall R.\alpha} \; Nec$$

is sound and complete for $\mathcal{ALC}$ semantics. In fact, by Lemmas 1 and 2, the Axiom 2.1 and this necessitation rule are an alternative axiomatization for $\mathcal{ALC}$.

Lemma 1 *The necessitation rule is a derived rule in the above Axiomatization.*

Proof Let α be a tautology, so that, from $\mathcal{ALC}$ semantics, $\alpha \equiv \top$ and hence, $\forall R.\alpha \equiv \forall R.\top$. Now, from the Axiom $\forall R.\top \equiv \top$ and we can conclude that $\forall R.\alpha \equiv \top$, that is, $\forall R.\alpha$ is also a tautology, and so it is provable by completeness. $\qquad \square$

Lemma 2 *The Axiom $\forall R.\top \equiv \top$ is derived from the necessitation rule.*

Proof If the necessitation rule is valid then whenever its premise is valid, its conclusion is valid. $\top$ is provable, so we can conclude that $\forall R.\top$ is also provable by the necessitation rule, so the equivalence $\forall R.\top \equiv \top$ is also provable.

Finally, we can state two useful facts following directly from the $\mathcal{ALC}$ semantics. Those facts will be used during this book to prove the soundness of the presented deduction systems. □

Fact 1 *If $C \sqsubseteq D$ then $\exists R.C \sqsubseteq \exists R.D$.*

Fact 2 *If $C \sqsubseteq D$ then $\forall R.C \sqsubseteq \forall R.D$.*

References

1. Baader, F.: The Description Logic Handbook: Theory, Implementation, and Applications. Cambridge University Press, Cambridge (2003)
2. Levesque, H., Brachman, R.: Expressiveness and tractability in knowledge representation and reasoning. Computat. Intell. **3**(2), 78–93 (1987)
3. Schild, K.: A correspondence theory for terminological logics: preliminary report. In: IJCAI'91: Proceedings of the 12th International Joint Conference on Artificial Intelligence, pp. 466–471. Morgan Kaufmann Publishers Inc., San Francisco, (1991). Also published at TR 91, Technische Universitat Berlin
4. Schmidt-Schauß, M., Smolka, G.: Attributive concept descriptions with complements. Artif. Intell. **48**(1), 1–26 (1991)

Chapter 3
The Sequent Calculus for $\mathcal{ALC}$

Abstract In this chapter, we present our first deduction system for $\mathcal{ALC}$, a Sequent Calculus for ACL named SC$_{\mathcal{ALC}}$. This first system is a labeled style proof system pure propositional in a sense that we don't deal with variables or individuals. Besides presenting the system, we also present the main proof theoretical properties of it. That is, we proof that the system is complete, sound and its cut rule can be removed without compromising its completeness.

Keywords Sequent calculus · Completeness theorem · Soundness theorem · Cut-elimination · Gentzen's Hauptsatz

3.1 A Sequent Calculus for $\mathcal{ALC}$

The Sequent Calculus for $\mathcal{ALC}$ that it is shown in Figs. 3.1 and 3.2 considers the extension of the language $\mathcal{ALC}$ presented in Sect. 2.1 for labeled concepts. The labels are a list of existencial or universal quantified roles names. Its syntax is as following:

$$L \to \forall R, \quad L \mid \exists R, \quad L \mid \emptyset$$
$$\phi_{lc} \to {}^{L}\phi_{c}$$

where R stands for atomic role names, L for list of labels and ϕ_c for $\mathcal{ALC}$ concept descriptions defined in Sect. 2.1.

Each labeled $\mathcal{ALC}$ concept has a straightforward $\mathcal{ALC}$ concept equivalent. For example, the $\mathcal{ALC}$ concept $\exists R_2.\forall Q_2.\exists R_1.\forall Q_1.\alpha$ has the same semantics of the labeled concept ${}^{\exists R_2,\forall Q_2,\exists R_1,\forall Q_1}\alpha$.

In other words, the list of labels is just the roles prefix of a concept. Labels are syntactic artifacts of our system, which means that labeled concepts and its equivalent $\mathcal{ALC}$ have the same semantics. The system was designed to be extended

A. Rademaker, *A Proof Theory for Description Logics*,
SpringerBriefs in Computer Science, DOI: 10.1007/978-1-4471-4002-3_3,
© The Author(s) 2012

$$\overline{\alpha \Rightarrow \alpha} \qquad\qquad\qquad \overline{\bot \Rightarrow \alpha}$$

$$\frac{\Delta \Rightarrow \Gamma}{\Delta, \delta \Rightarrow \Gamma}\ \text{weak-l} \qquad\qquad\qquad \frac{\Delta \Rightarrow \Gamma}{\Delta \Rightarrow \Gamma, \gamma}\ \text{weak-r}$$

$$\frac{\Delta, \delta, \delta \Rightarrow \Gamma}{\Delta, \delta \Rightarrow \Gamma}\ \text{contraction-l} \qquad\qquad\qquad \frac{\Delta \Rightarrow \Gamma, \gamma, \gamma}{\Delta \Rightarrow \Gamma, \gamma}\ \text{contraction-r}$$

$$\frac{\Delta_1, \delta_1, \delta_2, \Delta_2 \Rightarrow \Gamma}{\Delta_1, \delta_2, \delta_1, \Delta_2 \Rightarrow \Gamma}\ \text{perm-l} \qquad\qquad\qquad \frac{\Delta \Rightarrow \Gamma_1, \gamma_1, \gamma_2, \Gamma_2}{\Delta \Rightarrow \Gamma_1, \gamma_2, \gamma_1, \Gamma_2}\ \text{perm-r}$$

$$\frac{\Delta_1 \Rightarrow \Gamma_1, {}^L\alpha \qquad {}^L\alpha, \Delta_2 \Rightarrow \Gamma_2}{\Delta_1, \Delta_2 \Rightarrow \Gamma_1, \Gamma_2}\ \text{cut}$$

Fig. 3.1 The system SC$_{\mathcal{ALC}}$: structural rules

$$\frac{\Delta, {}^{L,\forall R}\alpha \Rightarrow \Gamma}{\Delta, {}^L(\forall R.\alpha)L_2 \Rightarrow \Gamma}\ \forall\text{-l} \qquad\qquad \frac{\Delta \Rightarrow \Gamma, {}^{L,\forall R}\alpha}{\Delta \Rightarrow \Gamma, {}^L(\forall R.\alpha)}\ \forall\text{-r}$$

$$\frac{\Delta, {}^{L,\exists R}\alpha \Rightarrow \Gamma}{\Delta, {}^L(\exists R.\alpha) \Rightarrow \Gamma}\ \exists\text{-l} \qquad\qquad \frac{\Delta \Rightarrow \Gamma, {}^{L,\exists R}\alpha}{\Delta \Rightarrow \Gamma, {}^L(\exists R.\alpha)}\ \exists\text{-r}$$

$$\frac{\Delta, {}^{\forall L}\alpha, {}^{\forall L}\beta \Rightarrow \Gamma}{\Delta, {}^{\forall L}(\alpha \sqcap \beta) \Rightarrow \Gamma}\ \sqcap\text{-l} \qquad\qquad \frac{\Delta \Rightarrow \Gamma, {}^{\forall L}\alpha \qquad \Delta \Rightarrow \Gamma, {}^{\forall L}\beta}{\Delta \Rightarrow \Gamma, {}^{\forall L}(\alpha \sqcap \beta)}\ \sqcap\text{-r}$$

$$\frac{\Delta, {}^{\exists L}\alpha \Rightarrow \Gamma \qquad \Delta, {}^{\exists L}\beta \Rightarrow \Gamma}{\Delta, {}^{\exists L}(\alpha \sqcup \beta) \Rightarrow \Gamma}\ \sqcup\text{-l} \qquad\qquad \frac{\Delta \Rightarrow \Gamma, {}^{\exists L}\alpha, {}^{\exists L}\beta}{\Delta \Rightarrow \Gamma, {}^{\exists L}(\alpha \sqcup \beta)}\ \sqcup\text{-r}$$

$$\frac{\Delta \Rightarrow \Gamma, {}^{\neg L}\alpha}{\Delta, {}^L\neg\alpha \Rightarrow \Gamma}\ \neg\text{-l} \qquad\qquad \frac{\Delta, {}^{\neg L}\alpha \Rightarrow \Gamma}{\Delta \Rightarrow \Gamma, {}^L\neg\alpha}\ \neg\text{-r}$$

$$\frac{\delta \Rightarrow \Gamma}{{}^{+\exists R}\delta \Rightarrow {}^{+\exists R}\Gamma}\ \text{prom-}\exists \qquad\qquad \frac{\Delta \Rightarrow \gamma}{{}^{+\forall R}\Delta \Rightarrow {}^{+\forall R}\gamma}\ \text{prom-}\forall$$

Fig. 3.2 The system SC$_{\mathcal{ALC}}$: logical rules

to description logics with role constructors and subsumptions of roles. This is one of the main reasons to use roles-as-labels in its formulation. Besides that, whenever roles are promoted to labels the rules of the calculus can compose or decompose concept description preserving role prefix stored as labels. In that way, labels are a kind of "context" where concept manipulation occurs.

Given that any labeled concept has an equivalent $\mathcal{ALC}$ concept, the semantics of a labeled concept can be given with the support of a formal transformation of labeled concepts into $\mathcal{ALC}$ concepts. We defined the function $\sigma : \phi_{lc} \rightarrow \phi_c$ that takes a labeled $\mathcal{ALC}$ concept an returns a $\mathcal{ALC}$ concept. Considering α an $\mathcal{ALC}$ concept description, the function σ is recursively defined as:

$$\sigma\left({}^{\emptyset}\alpha\right) = \alpha$$

$$\sigma\left({}^{\forall R, L}\alpha\right) = \forall R.\sigma\left({}^{L}\alpha\right)$$

$$\sigma\left({}^{\exists R, L}\alpha\right) = \exists R.\sigma\left({}^{L}\alpha\right)$$

Given σ, the semantics of a labeled concept γ is given by $\sigma\left(\gamma\right)^{\mathcal{I}}$.

We define $\Delta \Rightarrow \Gamma$ as a *sequent* where Δ and Γ are finite sequences of labeled concepts. The natural interpretation of the sequent $\Delta \Rightarrow \Gamma$ is the $\mathcal{ALC}$ formula:

$$\prod_{\delta \in \Delta} \sigma(\delta) \sqsubseteq \bigsqcup_{\gamma \in \Gamma} \sigma(\gamma)$$

The $SC_{\mathcal{ALC}}$ system is presented in Figs. 3.1 and 3.2. In all rules of the figures, the greek letters α and β stand for $\mathcal{ALC}$ concepts (formulas without labels), γ_i and δ_i stand for labeled concepts, Γ_i and Δ_i for lists of labeled concepts. For a clean presentation, the lists of labels are omitted whenever they are not used in the rule, this is the case of all structural rules in Fig. 3.1. The notation ${}^{L}\Gamma$ has to be taken as a list of labeled formulas of the form ${}^{L}\gamma_1, \ldots, {}^{L}\gamma_k$ for all $\gamma \in \Gamma$. The notation ${}^{+\forall R}\gamma$ (resp. ${}^{+\exists R}\gamma$) which can also be used with list of labeled concepts, ${}^{+\forall R}\Gamma$ (resp. ${}^{+\exists R}\Gamma$), means the addition of a label, $\forall R$ or $\exists R$ of a given role R, in front of the list of labels of γ, respectively in all $\gamma \in \Gamma$. Finally, we write ${}^{\exists L}\alpha$ to denote that all labels of L are existential quantificated and ${}^{\forall L}\alpha$ whenever all labels are universal quantificated (value restricted).

Considering the labeled formula ${}^{L}\alpha$, the notation ${}^{\neg L}\beta$ denotes exchanging the universal roles occurring in L for existential roles and vice-versa in a consistent way. Thus, if $\beta \equiv \neg\alpha$ then the formulas will be a negation each other. For example, ${}^{\neg\forall R, \exists Q}\beta$ is ${}^{\exists R, \forall Q}\neg\alpha$.

The system ought to be used by applying propositional rules, then the introduction of labels and then the quantification rules. This procedure will derive a normal derivation. Example 1 was taken from [1] and is useful to give an idea of how the rules of the $SC_{\mathcal{ALC}}$ system can be used.

Example 1 Given the $\mathcal{ALC}$ subsumption axiom:

$$\exists child.\top \sqcap \forall child.\neg(\exists child.\neg Doctor) \sqsubseteq \exists child.\forall child.Doctor \quad (3.1)$$

In $SC_{\mathcal{ALC}}$, we can prove that it is a theorem with the proof:

$$\cfrac{\cfrac{\cfrac{\cfrac{\cfrac{\cfrac{Doctor \Rightarrow Doctor}{^{\forall child}Doctor \Rightarrow {}^{\forall child}Doctor}\text{ prom-}\forall}{\top, {}^{\forall child}Doctor \Rightarrow {}^{\forall child}Doctor}\text{ weak-l}}{\top \Rightarrow {}^{\exists child}\neg Doctor, {}^{\forall child}Doctor}\text{ ¬-r}}{\top \Rightarrow {}^{\exists child}.\neg Doctor, {}^{\forall child}Doctor}\text{ ∃-r}}{{}^{\exists child}\top \Rightarrow {}^{\exists child}({}^{\exists child}.\neg Doctor), {}^{\exists child,\forall child}Doctor}\text{ prom-∃}}{\ }\text{ ¬-l}$$

$$\cfrac{\cfrac{\cfrac{\cfrac{\cfrac{{}^{\exists child}\top, {}^{\forall child}\neg({}^{\exists child}.\neg Doctor) \Rightarrow {}^{\exists child,\forall child}Doctor}{{}^{\exists child}\top, {}^{\forall child}\neg({}^{\exists child}.\neg Doctor) \Rightarrow {}^{\exists child}\forall child.Doctor}\text{ ∨-l}}{{}^{\exists child}\top, {}^{\forall child}\neg({}^{\exists child}.\neg Doctor) \Rightarrow {}^{\exists child}.\forall child.Doctor}\text{ ∃-r}}{{}^{\exists child}\top, \forall child.\neg({}^{\exists child}.\neg Doctor) \Rightarrow {}^{\exists child}.\forall child.Doctor}\text{ ∨-l}}{{}^{\exists child}.\top, \forall child.\neg({}^{\exists child}.\neg Doctor) \Rightarrow {}^{\exists child}.\forall child.Doctor}\text{ ∃-l}}{{}^{\exists child}.\top \sqcap \forall child.\neg({}^{\exists child}.\neg Doctor) \Rightarrow {}^{\exists child}.\forall child.Doctor}\text{ ⊓-l}$$

3.2 SC$_{\mathcal{ALC}}$ Soundness

The soundness of SC$_{\mathcal{ALC}}$ is proved by taking into account the intuitive meaning of each sequent and establishing that the truth preservation holds. From Sect. 3.1, a sequent $\Delta \Rightarrow \Gamma$ is equivalent in meaning to the $\mathcal{ALC}$ formula:

$$\prod_{\delta \in \Delta} \sigma(\delta) \sqsubseteq \bigsqcup_{\gamma \in \Gamma} \sigma(\gamma)$$

A sequent is defined to be *valid* or a *tautology* if and only if its corresponding $\mathcal{ALC}$ formula is.

When using the calculus, the usual axioms of a particular DL theory (TBox or an ontology) of the form $C \sqsubseteq D$ should be taken as sequents $C \Rightarrow D$. Labeled formulas occur only during the proof procedure, since they are in practical terms taken as intermediate data.

Theorem 1 (SC$_{\mathcal{ALC}}$ is sound) *Considering Ω a set of sequents, a theory or a TBox, let a Ω-proof be any SC$_{\mathcal{ALC}}$ proof in which sequents from Ω are permitted as initial sequents (in addition to the logical axioms). The soundness of SC$_{\mathcal{ALC}}$ states that if a sequent $\Delta \Rightarrow \Gamma$ has a Ω-proof, then $\Delta \Rightarrow \Gamma$ is satisfied by every interpretation which satisfies Ω. That is,*

$$\text{if } \Omega \vdash_{\text{SC}_{\mathcal{ALC}}} \Delta \Rightarrow \Gamma \text{ then } \Omega \models \prod_{\delta \in \Delta} \sigma(\delta) \sqsubseteq \bigsqcup_{\gamma \in \Gamma} \sigma(\gamma)$$

In the proof of Theorem 1 we will write $\Delta^{\mathcal{I}}$ as an abbreviation for the set interpretation of the conjunction of concepts in Δ, that is, $\bigcap_{\delta \in \Delta} \sigma(\delta)^{\mathcal{I}}$, and $\Gamma^{\mathcal{I}}$ as an abbreviation for the set interpretation of the disjunction of the concepts in Γ, $\bigcup_{\gamma \in \Gamma} \sigma(\gamma)^{\mathcal{I}}$.

During the proof below, we will use many times the axioms and facts from Sect. 2.6.

Proof We proof Theorem 1 by induction on the length of the Ω-proofs. The length of a Ω-proof is the number of applications for any derivation rule of the calculus in a top-down approach.

Base case

Proofs with length zero are proofs $\Omega \vdash \Delta \Rightarrow \Gamma$ where $\Delta \Rightarrow \Gamma$ occurs in Ω. In that case, it is easy to see that the theorem holds.

For the initial sequents, logical axioms like $C \Rightarrow C$, it is easy to see that $\sigma(C)^{\mathcal{I}} \subseteq \sigma(C)^{\mathcal{I}}$ for every interpretation $\mathcal{I}$ since every set is a subset of itself.

Induction hypothesis

As inductive hypothesis, we will consider that for proofs of length n the theorem holds. It is now sufficient to show that each of the derivation rules preserves the truth. That is, if the premises holds, the conclusion must also hold.

Cut rule

Given the sequents $\Delta_1 \Rightarrow \Gamma_1, {}^L C$ and ${}^L C, \Delta_2 \Rightarrow \Gamma_2$ then, by hypothesis, we know that they are valid and so

$$\bigcap_{\delta \in \Delta_1} \sigma(\delta)^{\mathcal{I}} \subseteq \bigcup_{\gamma \in \Gamma_1} \sigma(\gamma)^{\mathcal{I}} \cup \sigma({}^L C)^{\mathcal{I}}$$

and

$$\sigma({}^L C)^{\mathcal{I}} \cap \bigcap_{\delta \in \Delta_2} \sigma(\delta)^{\mathcal{I}} \subseteq \bigcup_{\gamma \in \Gamma_2} \sigma(\gamma)^{\mathcal{I}}$$

Let $\Delta_1^{\mathcal{I}} = \bigcap_{\delta \in \Delta_1} \sigma(\delta)^{\mathcal{I}}$, $\Gamma_1^{\mathcal{I}} = \bigcup_{\gamma \in \Gamma_1} \sigma(\gamma)^{\mathcal{I}}$, $\Delta_2^{\mathcal{I}} = \bigcap_{\delta \in \Delta_2} \sigma(\delta)^{\mathcal{I}}$, $\Gamma_2^{\mathcal{I}} = \bigcup_{\gamma \in \Gamma_2} \sigma(\gamma)^{\mathcal{I}}$ and $X = \sigma({}^L C)^{\mathcal{I}}$. Now me must show that the application of the *cut* rule preserves the set inclusion. In other words, given $\Delta_1^{\mathcal{I}} \subseteq (\Gamma_1^{\mathcal{I}} \cup X)$ and $(X \cap \Delta_2^{\mathcal{I}}) \subseteq \Gamma_2^{\mathcal{I}}$, we must have $(\Delta_1^{\mathcal{I}} \cap \Delta_2^{\mathcal{I}}) \subseteq (\Gamma_1^{\mathcal{I}} \cup \Gamma_2^{\mathcal{I}})$. What is easy to show using the standard set theory.

Rules *weak-l* and *weak-r*

Given the sequent $\Delta \Rightarrow \Gamma$, by the inductive hypothesis we know that

$$\Delta^{\mathcal{I}} \subseteq \Gamma^{\mathcal{I}}$$

By set theory, $\Delta^{\mathcal{I}} \cap X \subseteq \Gamma^{\mathcal{I}}$ and $\Delta^{\mathcal{I}} \subseteq \Gamma^{\mathcal{I}} \cup X$ for any set X interpretation of a labeled concept α. In the first case, we have the interpretation of $\Delta, \alpha \Rightarrow \Gamma$. In the second case, we have the interpretation of $\Delta \Rightarrow \Gamma, \alpha$. This is sufficient to show the soundness of both rules.

Rules *perm-l* and *perm-r*

By the definition of the meaning of a sequent and its semantics, it is easy to see that both rules are sound. Note that the order of the formulas in both sides of a sequent does not change the sequent semantics.

Rules *prom-∀* and *prom-∃*

The soundness of rule prom-∃ if easily proved using the Fact 1 and Axiom 2.3. The soundness of rule prom-∀ is proved using Fact 2 and the Axiom 2.1.

Rules ∀-r, ∀-l, ∃-r and ∃-l

From the definition of σ function, we know that in all those four rules, both the premises and the conclusions have, given a interpretation function, the same semantics.

Rules ⊓-l and ⊓-r

In order to prove the soundness of those rules we need the $\mathcal{ALC}$ Axiom 2.1 that states the distributivity of the universal quantified constructor over the conjunction. Moreover, we must observe that both rules have an important proviso. That is, they are restricted to details only with labeled concepts were all labels are universal quantified. This restriction permits us to apply the Axiom 2.1 inductively.

Taking the sequent $\Delta, {}^{L}\alpha, {}^{L}\beta \Rightarrow \Gamma$ valid as hypothesis, we have:

$$\left(\Delta^{\mathcal{I}} \cap \sigma({}^{L}\alpha)^{\mathcal{I}} \cap \sigma({}^{L}\beta)^{\mathcal{I}} \right) \subseteq \Gamma^{\mathcal{I}}$$

To show that the rule (⊓-l) is sound, We must prove that $\Delta, {}^{L}(\alpha \sqcap \beta) \Rightarrow \Gamma$ is also valid. In other words, that $\Delta^{\mathcal{I}} \cap \sigma({}^{L}(\alpha \sqcap \beta))^{\mathcal{I}} \subseteq \Gamma^{\mathcal{I}}$ holds. What is true by the definition of σ, the Axiom 2.1 and the rule proviso which allows us to apply the Axiom 2.1 over the list of labels L.

Now consider the rule (⊓-r). By induction hypothesis, the sequents $\Delta \Rightarrow \Gamma, {}^{L}\alpha$ and $\Delta \Rightarrow \Gamma, {}^{L}\beta$ are valid, and so,

$$\Delta^{\mathcal{I}} \subseteq \Gamma^{\mathcal{I}} \cup \sigma({}^{L}\alpha)^{\mathcal{I}} \quad and \quad \Delta^{\mathcal{I}} \subseteq \Gamma^{\mathcal{I}} \cup \sigma({}^{L}\beta)^{\mathcal{I}}$$

holds for all interpretations $.^{\mathcal{I}}$. Now, suppose the application of the rule (⊓-r) over the two sequents above. We must show that $\Delta \Rightarrow \Gamma, {}^{L}(\alpha \sqcap \beta)$ is also valid, that is,

$$\Delta^{\mathcal{I}} \subseteq \Gamma^{\mathcal{I}} \cup \sigma({}^{L}(\alpha \sqcap \beta))^{\mathcal{I}}$$

holds. But by basic set theory we have

$$\Delta^{\mathcal{I}} \subseteq ((\Gamma^{\mathcal{I}} \cup \sigma(^{L}\alpha)^{\mathcal{I}}) \cap (\Gamma^{\mathcal{I}} \cup \sigma(^{L}\beta)^{\mathcal{I}}))$$

And by distributive law

$$\Delta^{\mathcal{I}} \subseteq \Gamma^{\mathcal{I}} \cup (\sigma(^{L}\alpha)^{\mathcal{I}} \cap \sigma(^{L}\beta)^{\mathcal{I}})$$

Finally, by the definition of σ, the Axiom 2.1 and the rule proviso, we can conclude that the rule conclusion if valid.

Rules ⊔-l and ⊔-r

In both rules the proviso is that the labels list L must contain only existential quantified roles. The soundness of both rules is proved with the support of this proviso and the Axiom 2.3, applied inductively over the labels lists.

As inductive hypothesis the sequents $\Delta, {}^{L}\alpha \Rightarrow \Gamma$ and $\Delta, {}^{L}\beta \Rightarrow \Gamma$ are valid. That is, given $\Delta^{\mathcal{I}} = \bigcap_{\delta \in \Delta} \sigma(\delta)^{\mathcal{I}}$ and $\Gamma^{\mathcal{I}} = \bigcup_{\gamma \in \Gamma} \sigma(\gamma)^{\mathcal{I}}$, we know that

$$\Delta^{\mathcal{I}} \cap \sigma(^{L}\alpha)^{\mathcal{I}} \subseteq \Gamma^{\mathcal{I}} \quad \text{and} \quad \Delta^{\mathcal{I}} \cap \sigma(^{L}\beta)^{\mathcal{I}} \subseteq \Gamma^{\mathcal{I}}$$

holds. Now considering the application of the rule (⊔-l) over the two sequents above we must prove that the resulting sequent $\Delta, {}^{L}(\alpha \sqcup \beta) \Rightarrow \Gamma$ is also valid:

$$\Delta^{\mathcal{I}} \cap \sigma(^{\emptyset}(\alpha \sqcup \beta)^{L})^{\mathcal{I}} \subseteq \Gamma^{\mathcal{I}}$$

Following from the hypothesis and basic set theory we know that if $\Delta^{\mathcal{I}} \cap X_1 \subseteq \Gamma^{\mathcal{I}}$ and $\Delta^{\mathcal{I}} \cap X_2 \subseteq \Gamma^{\mathcal{I}}$ than $(\Delta^{\mathcal{I}} \cap X_1) \cup (\Delta^{\mathcal{I}} \cap X_2) \subseteq \Gamma^{\mathcal{I}}$ what gives us

$$\Delta^{\mathcal{I}} \cap (\sigma(^{L}\alpha)^{\mathcal{I}} \cup \sigma(^{L}\beta)^{\mathcal{I}}) \subseteq \Gamma^{\mathcal{I}}$$

and by the Axiom 2.3 applied inductively over the list L we have the desired semantics of the resulting sequent:

$$\Delta^{\mathcal{I}} \cap (\sigma(^{L}(\alpha \sqcup \beta))^{\mathcal{I}}) \subseteq \Gamma^{\mathcal{I}}$$

For rule (⊔r) the inductive hypothesis is that $\Delta \Rightarrow \Gamma, {}^{L}\alpha, {}^{L}\beta$ is valid. And so, the following statement must hold:

$$\Delta^{\mathcal{I}} \subseteq \Gamma^{\mathcal{I}} \cup \sigma(^{L}\alpha)^{\mathcal{I}} \cup \sigma(^{L}\beta)^{\mathcal{I}}$$

Now by the Axiom 2.3 we can rewrite to

$$\Delta^{\mathcal{I}} \subseteq \Gamma^{\mathcal{I}} \cup \sigma(^{L}(\alpha \sqcup \beta))^{\mathcal{I}}$$

What is the semantics of the rule conclusion.

Rules ¬-l and ¬-r

Given a concept $^L\alpha$ and a interpretation $.^\mathcal{I}$ we define the set $X = \sigma(^L\alpha)^\mathcal{I}$ and the interpretation of its negation, $\sigma(^{\neg L}\neg\alpha)^\mathcal{I}$, will be the set $\overline{X} = \Delta^\mathcal{I} \setminus X$.

For rule (¬-l), the inductive hypothesis is that the premise $\Delta \Rightarrow \Gamma, {}^L\alpha$ is valid. Which means that $\Delta^\mathcal{I} \subseteq (\Gamma^\mathcal{I} \cup X)$. From the basic set theory this implies that $(\Delta^\mathcal{I} \cap \overline{X}) \subseteq \Gamma^\mathcal{I}$, which is the interpretation of the conclusion.

For rule (¬-r), the inductive hypothesis is that the premise $\Delta, {}^L\alpha \Rightarrow \Gamma$ is valid. Which means that $(\Delta^\mathcal{I} \cap X) \subseteq \Gamma^\mathcal{I}$. From the basic set theory this implies that $\Delta^\mathcal{I} \subseteq (\Gamma^\mathcal{I} \cup \overline{X})$, the interpretation of the conclusion as desired.   □

3.3 The Completeness of $\mathrm{SC}_{\mathcal{ALC}}$

We show the relative completeness of $\mathrm{SC}_{\mathcal{ALC}}$ regarding the axiomatic presentation of $\mathcal{ALC}$ from Sect. 2.6. Since $\mathcal{ALC}$ formulas are not labeled, the completeness must take into account only formulas with empty list of labels. Proceeding in this way, the $\mathcal{ALC}$ sequent calculus deduction rules without labels behave exactly as the sequent calculus rules for classical propositional logic. Thus, in order to prove that $\mathrm{SC}_{\mathcal{ALC}}$ is complete, we only have to derive the axioms above.

The derivation of the rule of necessitation is accomplished by

$$\frac{\dfrac{\Rightarrow \alpha}{\Rightarrow {}^{\forall R}\alpha}\ \text{prom-}\forall}{\Rightarrow \forall R.\alpha}\ \text{∀-r}$$

The derivation of the Axiom 2.1 is obtained from the following derivations. First we consider the case:

$$\forall R.(\alpha \sqcap \beta) \sqsubseteq \forall R.\alpha \sqcap \forall R.\beta$$

$$\frac{\dfrac{\dfrac{\dfrac{{}^{\forall R}\alpha \Rightarrow {}^{\forall R}\alpha}{{}^{\forall R}\alpha, {}^{\forall R}\beta \Rightarrow {}^{\forall R}\alpha}\ \text{weak-l}}{{}^{\forall R}\alpha, {}^{\forall R}\beta \Rightarrow \forall R.\alpha}\ \text{∀-r} \quad \dfrac{\dfrac{{}^{\forall R}\beta \Rightarrow {}^{\forall R}\beta}{{}^{\forall R}\alpha, {}^{\forall R}\beta \Rightarrow {}^{\forall R}\beta}\ \text{weak-l}}{{}^{\forall R}\alpha, {}^{\forall R}\beta \Rightarrow \forall R.\beta}\ \text{∀-r}}{\dfrac{\dfrac{{}^{\forall R}\alpha, {}^{\forall R}\beta \Rightarrow \forall R.\alpha \sqcap \forall R.\beta}{{}^{\forall R}(\alpha \sqcap \beta) \Rightarrow \forall R.\alpha \sqcap \forall R.\beta}\ \text{⊓-l}}{\forall R.(\alpha \sqcap \beta) \Rightarrow \forall R.\alpha \sqcap \forall R.\beta}\ \text{∀-l}}\ \text{⊓-r}$$

Finally, we prove the subsumption from right to left:

$$\forall R.\alpha \sqcap \forall R.\beta \sqsubseteq \forall R.(\alpha \sqcap \beta)$$

by

$$
\cfrac{
\cfrac{
\cfrac{
\cfrac{{}^{\forall R}\alpha \Rightarrow {}^{\forall R}\alpha}{\forall R.\beta, {}^{\forall R}\alpha \Rightarrow {}^{\forall R}\alpha}\text{ weak-l}
}{\forall R.\beta, \forall R.\alpha \Rightarrow {}^{\forall R}\alpha}\text{ ∨-l}
\qquad
\cfrac{
\cfrac{
\cfrac{{}^{\forall R}\beta \Rightarrow {}^{\forall R}\beta}{\forall R.\alpha, {}^{\forall R}\beta \Rightarrow {}^{\forall R}\beta}\text{ weak-l}
}{\forall R.\alpha, \forall R.\beta \Rightarrow {}^{\forall R}\beta}\text{ ∨-l}
}{\forall R.\alpha, \forall R.\beta \Rightarrow {}^{\forall R}(\alpha \sqcap \beta)}\text{ ⊓-r}
}{\forall R.\alpha \sqcap \forall R.\beta \Rightarrow {}^{\forall R}(\alpha \sqcap \beta)}\text{ ⊓-l}
$$

$$
\cfrac{\forall R.\alpha \sqcap \forall R.\beta \Rightarrow {}^{\forall R}(\alpha \sqcap \beta)}{\forall R.\alpha \sqcap \forall R.\beta \Rightarrow \forall R.(\alpha \sqcap \beta)}\text{ ∨-r}
$$

3.4 The Cut-Elimination Theorem

In this section we adopt the usual terminology of proof theory for sequent calculus presented in [2, 3]. We follow Gentzen's original proof for cut elimination with the introduction of the *mix* rule.

Let δ be a labeled formula. An inference of the following form is called *mix* with respect to ψ, a labeled concept:

$$
\frac{\Delta_1 \Rightarrow \Gamma_1 \quad \Delta_2 \Rightarrow \Gamma_2}{\Delta_1, \Delta_2^* \Rightarrow \Gamma_1^*, \Gamma_2} \ (\psi)
$$

where both Γ_1 and Δ_2 contain the formula δ, and Γ_1^* and Δ_2^* are obtained from Γ_1 and Δ_2 respectively by deleting all the occurrences of δ in them.

But in order to obtain an easier presentation of our cut elimination we introduce four additional rules of inference called *quasi-mix* rules.

$$
\frac{{}^L\delta \Rightarrow \Gamma_1 \quad \Delta_2 \Rightarrow \Gamma_2}{{}^{\exists R,L}\delta, \Delta_2^* \Rightarrow {}^{+\exists R}\Gamma_1^*, \Gamma_2} \ ({}^L\alpha, {}^{\exists R,L}\alpha)
\qquad
\frac{\Delta_1 \Rightarrow \Gamma_1 \quad {}^L\alpha \Rightarrow \Gamma_2}{\Delta_1 \Rightarrow \Gamma_1^*, {}^{+\exists R}\Gamma_2} \ ({}^{\exists R,L}\alpha, {}^L\alpha)
$$

$$
\frac{\Delta_1 \Rightarrow {}^L\alpha \quad \Delta_2 \Rightarrow \Gamma_2}{{}^{+\forall R}\Delta_1, \Delta_2^* \Rightarrow \Gamma_2} \ ({}^L\alpha, {}^{\forall R,L}\alpha)
\qquad
\frac{\Delta_1 \Rightarrow \Gamma_1 \quad \Delta_2 \Rightarrow {}^L\gamma}{\Delta_1, {}^{+\forall R}\Delta_2^* \Rightarrow \Gamma_1^*, {}^{\forall R,L}\gamma} \ ({}^{\forall R,L}\alpha, {}^L\alpha)
$$

where in each rule, the tuple of concepts on the right indicates the two mix formulas of this inference rule. Γ_1, the list of formulas on the right from the left-side premise contains the first projection of the tuple, Δ_2, the list of formulas on the left from the right-side premisse, contains the second projection. Γ_1^* and Δ_2^* are obtained from Γ_1 and Δ_2 by deleting all occurrences of the first and second tuple's projection, respectively. The notation ${}^{+\exists R}\Delta$ (resp. ${}^{+\forall R}\Delta$) means the addition of $\exists R$ (resp. $\forall R$) on the list of labels of all $\delta \in \Delta$.

By the definitions of *mix* and *quasi-mix* rules, the *mix* rule is a special case of *quasi-mix* rules in which both mix formulas in the tuple are equal. Therefore, we can also consider a *quasi-mix* the *mix* rule.

Definition 9 *(The SC*$_{\mathcal{ALC}}$ system) We call SC*$_{\mathcal{ALC}}$ the new system obtained from SC$_{\mathcal{ALC}}$ by replacing the cut rule by the* quasi-mix *(and mix) rules.*

Lemma 3 *The systems SC$_{\mathcal{ALC}}$ and SC*$_{\mathcal{ALC}}$ are equivalent, that is, a sequent is SC$_{\mathcal{ALC}}$-provable if and only if that sequent is also SC*$_{\mathcal{ALC}}$-provable.*

Proof The four *quasi-mix* rules are derived from inferences where the promotional rules (prom-∀ and prom-∃) are applied just before a *mix* rule. In that way, one can transform all the applications of *quasi-mix* rule into a sequence of prom-∀ or prom-∃ followed by *mix* rules applications. All applications of *mix* rule can then be replaced by applications of *cut* rule provide that all the repetitions of the cut formula in the upper sequents being first transformed into just one occurrence on each sequent. This is easily done by one or more application of the contraction and permutation rules.

To illustrate the process, let us consider an application of a *quasi-mix* rule in the SC*$_{\mathcal{ALC}}$-proof fragment below where Π_n are proof fragments. The double-line labeled with "perm*; contract*" means the application of rule permutation one or more times followed by one or more applications of contraction rule.

$$
\begin{array}{cc}
\Pi_1 & \Pi_2 \\
\dfrac{\Delta_1 \Rightarrow \Gamma_1 \qquad {}^L\alpha \Rightarrow \Gamma_2}{\Delta_1 \Rightarrow \Gamma_1^*,\,{}^{+\exists R}\Gamma_2}\ {\scriptstyle(\exists R, {}^L\alpha, {}^L\alpha)} \\
\Pi_3
\end{array}
$$

And its corresponding SC$_{\mathcal{ALC}}$-proof:

$$
\begin{array}{cc}
\Pi_1 & \Pi_2 \\
\dfrac{\Delta_1 \Rightarrow \Gamma_1}{\Delta_1 \Rightarrow \Gamma_1^*,\,{}^{\exists R, {}^L\alpha}}\ {\scriptstyle \text{perm*; contract*}} \qquad & \dfrac{{}^L\alpha \Rightarrow \Gamma_2}{{}^{\exists R, {}^L}\alpha \Rightarrow {}^{+\exists R}\Gamma_2}\ {\scriptstyle \text{prom-}\exists} \\[2ex]
\multicolumn{2}{c}{\dfrac{\rule{0pt}{0pt}}{\Delta_1 \Rightarrow \Gamma_1^*,\,{}^{+\exists R}\Gamma_2}\ {\scriptstyle \text{cut}}} \\
\multicolumn{2}{c}{\Pi_3}
\end{array}
$$

□

By the proof of Lemma 3, a derivation Π of $\Delta \Rightarrow \Gamma$ in SC$_{\mathcal{ALC}}$ with cuts can be transformed in a derivation Π' of $\Delta \Rightarrow \Gamma$ in SC*$_{\mathcal{ALC}}$ with *quasi-mixes* (and *mixes*). So that, it is sufficient to show that the *quasi-mix* (and *mix*) rules are redundant in SC*$_{\mathcal{ALC}}$, since a proof in SC*$_{\mathcal{ALC}}$ without *quasi-mix* (and *mix*) is at the same time a proof in SC$_{\mathcal{ALC}}$ without cut.

Definition 10 *(SC^T$_{\mathcal{ALC}}$ system) SC$_{\mathcal{ALC}}$ was defined with initial sequents of the form $\alpha \Rightarrow \alpha$ with α a $\mathcal{ALC}$ concept definition (logical axiom). However, it is often convenient to allow for other initial sequents. So if T is a set of sequents of the form $\Delta \Rightarrow \Gamma$, where Δ and Γ are sequences of $\mathcal{ALC}$ concept descriptions(non-logical* *axioms), we define* $SC^{\mathcal{T}}{}_{\mathcal{ALC}}$ *to be the proof system defined like* $SC_{\mathcal{ALC}}$ *but allowing initial sequents to be from* $\mathcal{T}$ *too.*

The Definition 10 can be extended to the system $SC^*{}_{\mathcal{ALC}}$ in the same way, obtaining the systems $SC^{*\mathcal{T}}{}_{\mathcal{ALC}}$.

Definition 11 *(Free-quasi-mix free proof) Let P be an* $SC^{*\mathcal{T}}{}_{\mathcal{ALC}}$*-proof. A formula occurring in P is anchored (by an* $\mathcal{T}$*-sequent) if it is a direct descendent of a formula from* $\mathcal{T}$ *occurring in an initial sequent. A* quasi-mix *inference in P is anchored if either:*

(i) the mix formulas are not atomic and at least one of the occurrences of the mix formulas in the upper sequents is anchored, or
(ii) the mix formulas are atomic and both of the occurrences of the mix formulas in the upper sequents are anchored.

A quasi-mix *inference which is not anchored is said to be* free. *A proof P is* free-quasi-mix free *if it contains no free* quasi-mixes.

Given that a *mix* is a special quase of *quasi-mix*, the Definition 11 can also be used to define *free* mixes. If a proof *P* is *free-quasi-mix free* it is also *free-mix free*.

Theorem 2 (Free-quasi-mix Elimination) *Let* $\mathcal{T}$ *be a set of sequents. If* $SC^{*\mathcal{T}}{}_{\mathcal{ALC}} \vdash \Delta \Rightarrow \Gamma$ *then there is a* free-quasi-mix *free* $SC^{*\mathcal{T}}{}_{\mathcal{ALC}}$*-proof of* $\Delta \Rightarrow \Gamma$.

Theorem 2 is a consequence of the following lemma.

Lemma 4 *If P is a proof of S (in* $SC^{*\mathcal{T}}{}_{\mathcal{ALC}}$*) which contains only one* free-quasi-mix, *occurring as the last inference, then S is provable without any* free-quasi-mix.

Theorem 2 is obtained from Lemma 4 by simple induction over the number of *quasi-free-mix* occurring in a proof *P*.

We can now concentrate our attention on Lemma 4. First we define three scalars as a measure of the complexity of the proof. The *grade* of a formula $^L\alpha$ is defined as the number of logical symbols of α (denoted by $g(^L\alpha)$). The *label-degree* of a formula $^L\alpha$ is defined as $ld(^L\alpha) = |L|$ where $|L|$ means the length of the list L.

Let *P* be a proof containing only one *quasi-mix* as its last inference:

$$J \; \frac{\Delta_1 \Rightarrow \Gamma_1 \qquad \Delta_2 \Rightarrow \Gamma_2}{\Delta_1, \Delta_2^* \Rightarrow \Gamma_1^*, \Gamma_2} \; (\gamma, \gamma')$$

The grade of a *quasi-mix* is

$$g(\gamma, \gamma') = g(\gamma) + g(\gamma')$$

Given that, the grade of a *mix* (a special quase of *quasi-mix*) is the double of the grade of the *mix* formula.

In a similar way, the label-degree of a *quasi-mix* is

$$ld(\gamma, \gamma') = ld(\gamma) + ld(\gamma')$$

and the label-degree of a *mix* is again the double of the lable-degree of the *mix* formula.

We say that the grade of P (denote by $g(P)$) and the label-degree of P (denoted by $ld(P)$) are the grade and label-degree of that *quasi-mix*.

We refer to the left and right sequents as S_1 and S_2 respectively, and to the lower sequent as S. We call a thread in P a left (or right) thread if it contains the left (or right) upper sequent of the *quasi-mix* J. The *rank* of the thread $\mathcal{F}$ in P is defined as the number of consecutive sequents, counting upward from the left (right) upper sequent of J, that contains γ (γ') in its succedent (antecedent). Since the left (right) upper sequent always contains the mix formulas, the rank of a thread in P is at least 1. The rank of a thread $\mathcal{F}$ in P is denoted by $rank(\mathcal{F}; P)$ and is defined as follows:

$$rank_l(P) = \max_{\mathcal{F}}(rank(\mathcal{F}; P)),$$

where $\mathcal{F}$ ranges over all the left threads in P, and

$$rank_r(P) = \max_{\mathcal{F}}(rank(\mathcal{F}; P)),$$

where $\mathcal{F}$ ranges over all the right threads in P. The rank of P is defined as

$$rank(P) = rank_l(P) + rank_r(P),$$

where $rank(P) \geq 2$.

Proof We prove Lemma 4 by lexicographically induction on the ordered triple (grade, label-degree, rank) of the proof P. We divide the proof into two main cases, namely $rank = 2$ and $rank > 2$ (regardless of the grade and label-degree).

Case 1: $rank = 2$

We shall consider several cases according to the form of the proofs of the upper sequents of the *quasi-mix*.

(1.1) The left upper sequent S_1 is a logical initial sequent. There are several cases to be examined.

   (a) P has the form:

$$J \, \dfrac{\alpha \Rightarrow \alpha \qquad \overset{\textstyle P_1}{\Delta_2 \Rightarrow \Gamma_2}}{{}^{\exists R}\alpha, \Delta_2^* \Rightarrow \Gamma_2} \; {}_{(\alpha, \, \exists R\alpha)}$$

We can easily obtain the same end-sequent without using the *quasi-mix* as follows:[1]

$$
\cfrac{
  \cfrac{
    \cfrac{P_1}{\Delta_2 \Rightarrow \Gamma_2}\ \text{perm*}
  }{^{\exists R}\alpha, \ldots, {}^{\exists R}\alpha, \Delta_2^* \Rightarrow \Gamma_2}\ \text{contract*}
}{^{\exists R}\alpha, \Delta_2^* \Rightarrow \Gamma_2}
$$

All other cases, that it, other *quasi-mix* occurrences in a similar proof format, are treated in a similar way. Note also that a logical initial sequent can only have $\mathcal{ALC}$ formulas on both sides of the sequent.

(1.2) The right upper sequent S_2 is a logical initial sequent. Similar as Case 1.1 above.

(1.3) S_1 or S_2 (or both) are non-logical initial sequents. In this case, it is obvious that the *quasi-mix* is not a *free* and it will be not eliminated.

(1.4) Neither S_1 nor S_2 are initial sequents, and S_1 is the lower sequent of a structural inference J_1. Since $rank_l(P) = 1$, the mix formula ψ cannot appear in the succedent of the premisse of J_1, that is, J_1 must be the *weak-r* that introduced ψ. Again there are several cases to be examined for each possible *quasi-mix* rule used.

(a) Let us consider the *quasi-mix* case $(^L\alpha, {}^{\exists R,L}\alpha)$:

$$
J\ \cfrac{
  \cfrac{\cfrac{P_1}{\delta \Rightarrow \Gamma_1}}{\delta \Rightarrow \Gamma_1, {}^L\alpha}\ J_1 \qquad \cfrac{P_2}{\Delta_2 \Rightarrow \Gamma_2}
}{^{+\exists R}\delta, \Delta_2^* \Rightarrow {}^{+\exists R}\Gamma_1, \Gamma_2}\ (^L\alpha, {}^{\exists R,L}\alpha)
$$

where Γ_1 does not contain $^L\alpha$. We can eliminate the *quasi-mix* as follows:

$$
\cfrac{
  \cfrac{
    \cfrac{
      \cfrac{P_1}{\delta \Rightarrow \Gamma_1}\ \text{prom-1}
    }{^{+\exists R}\delta \Rightarrow {}^{+\exists R}\Gamma_1}
  }{\Delta_2^*, {}^{+\exists R}\delta \Rightarrow {}^{+\exists R}\Gamma_1, \Gamma_2}\ \text{weak*}
}{^{+\exists R}\delta, \Delta_2^* \Rightarrow {}^{+\exists R}\Gamma_1, \Gamma_2}\ \text{perm*}
$$

All other cases are treated in a similar way.

(1.5) The same conditions that hold for Case (1.4) but with S_2 as the lower sequent of structural inference instead of S_1. As in Case (1.4).

(1.6) Neither S_1 nor S_2 are initial sequents and S_1 is the lower sequent of a prom-∃ rule application and J is a *mix* rule application.

---

[1] The notation *contract** (*perm**) means zero or more applications of contraction (permutation) rule.

$$J \quad \frac{\dfrac{\dfrac{P_1}{\delta \Rightarrow \Gamma_1}}{^{+\exists R}\delta \Rightarrow {}^{+\exists R}\Gamma_1} \text{ prom-}\exists \qquad \dfrac{P_2}{\Delta_2 \Rightarrow \Gamma_2}}{^{+\exists R}\delta, \Delta_2^* \Rightarrow {}^{+\exists R}\Gamma_1^*, \Gamma_2} ({}^{+\exists R}\gamma)$$

where by assumption none of the proofs P_n for $n \in \{1, 2\}$ contain a mix or quasi-mix. Moreover, Γ_1 does not contain $^{+\exists R}\gamma$ since $rank_l(P) = 1$. That is, the prom-$\exists$ rule introduced the mix formula of J. We can replace the application of the *mix* rule by an application of *quasi-mix* rule as follows:

$$\frac{\dfrac{P_1}{\delta \Rightarrow \Gamma_1} \qquad \dfrac{P_2}{\Delta_2 \Rightarrow \Gamma_2}}{^{+\exists R}\delta, \Delta_2^* \Rightarrow {}^{+\exists R}\Gamma_1^*, \Gamma_2} (\gamma, {}^{+\exists R}\gamma)$$

The new *quasi-mix* rule has label-degree less than the label-degree of the original *mix* rule, $ld({}^{+\exists R}\gamma, {}^{+\exists R}\gamma)$. So by the induction hypothesis, we can obtain a proof which contains no mixes.

(1.7)   Similar case as above with S_1 being lower sequent of a prom-$\forall$ or S_2 being lower sequent of prom-$\exists$ or prom-$\forall$. We apply similar transformation of *mix* application into *quasi-mix* rules applications. Always "moving" the *mix* upward into the direction of the prom-$\forall$ or prom-$\exists$ inference.

(1.8)   Both S_1 and S_2 are lower sequents of logical inferences and $rank_l(P) = rank_r(P) = 1$, J being a *mix* with the mix formula γ of each side being the principal formula of the logical inference. We use induction on the grade, distinguishing several cases according to the outermost logical symbol of γ:

(i)   The outermost logical symbol is $\sqcap$. P has the form:

$$\frac{\dfrac{\dfrac{P_1}{\Delta_1 \Rightarrow \Gamma_1, {}^L\alpha} \quad \dfrac{P_2}{\Delta_1 \Rightarrow \Gamma_1, {}^L\beta}}{\Delta_1 \Rightarrow \Gamma_1, {}^L(\alpha \sqcap \beta)} \sqcap\text{-r} \qquad \dfrac{\dfrac{P_3}{\Delta_2, {}^L\alpha, {}^L\beta \Rightarrow \Gamma_2}}{\Delta_2, {}^L(\alpha \sqcap \beta) \Rightarrow \Gamma_2} \sqcap\text{-l}}{\Delta_1, \Delta_2 \Rightarrow \Gamma_1, \Gamma_2} ({}^L(\alpha \sqcap \beta))$$

where by assumption none of the proofs P_n for $n \in \{1, 2, 3\}$ contain a quasi-mix. We transform P into:

$$\frac{\dfrac{P_2}{\Delta_1 \Rightarrow \Gamma_1, {}^L\beta} \qquad \dfrac{\dfrac{P_1}{\Delta_1 \Rightarrow \Gamma_1, {}^L\alpha} \quad \dfrac{P_3}{\Delta_2, {}^L\alpha, {}^L\beta \Rightarrow \Gamma_2}}{\Delta_1, \Delta_2, {}^L\beta \Rightarrow \Gamma_1, \Gamma_2} ({}^L\alpha)}{\dfrac{\Delta_1, \Delta_1, \Delta_2 \Rightarrow \Gamma_1, \Gamma_1, \Gamma_2}{\Delta_1, \Delta_2 \Rightarrow \Gamma_1, \Gamma_2} \text{ perm*; contract*}} ({}^L\beta)$$

which contains two mixes but both with grade less than $g({}^L(\alpha \sqcap \beta))$. So by induction hypothesis, we can obtain a proof which contains no mixes. Note that the mix $({}^L\alpha)$ is now the last inference rule of a proof which contains no mix. Given that, this mix can be omitted using the transformations defined above.

(ii) The outermost logical symbol is $\sqcup$. In this case S_1 and S_2 must be lower sequents of $\sqcup$-r and $\sqcup$-l rule, respectively:

$$
\cfrac{
\cfrac{P_1}{\Delta_1 \Rightarrow \Gamma_1, {}^L\alpha, {}^L\beta}{\Delta_1 \Rightarrow \Gamma_1, {}^L(\alpha \sqcup \beta)} \;\sqcup\text{-r}
\qquad
\cfrac{\cfrac{P_2}{\Delta_2, {}^L\alpha \Rightarrow \Gamma_2} \quad \cfrac{P_3}{\Delta_2, {}^L\beta \Rightarrow \Gamma_2}}{\Delta_2, {}^L(\alpha \sqcup \beta) \Rightarrow \Gamma_2} \;\sqcup\text{-l}
}{\Delta_1, \Delta_2 \Rightarrow \Gamma_1, \Gamma_2} \;({}^L(\alpha \sqcup \beta))
$$

where, by hypothesis, none of the proofs P_n for $n \in \{1, 2, 3\}$ contain a *quasi-mix*. This proof can be transformed into:

$$
\cfrac{
\cfrac{\cfrac{P_1}{\Delta_1 \Rightarrow \Gamma_1, {}^L\alpha, {}^L\beta} \quad \cfrac{P_2}{\Delta_2, {}^L\alpha \Rightarrow \Gamma_2}}{\Delta_1, \Delta_2 \Rightarrow \Gamma_1, \Gamma_2, {}^L\beta} \;({}^L\alpha)
\qquad
\cfrac{P_3}{\Delta_2, {}^L\beta \Rightarrow \Gamma_2}
}{
\cfrac{\cfrac{\Delta_1, \Delta_2, \Delta_2 \Rightarrow \Gamma_1, \Gamma_2, \Gamma_2}{} \;({}^L\beta)}{\Delta_1, \Delta_2 \Rightarrow \Gamma_1, \Gamma_2} \;\text{perm*; contract*}
}
$$

This proof contains two mix applications, but both with grade less than $g({}^L(\alpha \sqcup \beta))$. So by the induction hypothesis, we can obtain a proof which contains no mixes. As mentioned above, the new created mixes are now the last inference rule of proofs which contains no mix.

(iii) The outermost logical symbol is $\forall$. In this case S_1 and S_2 must be lower sequents of $\forall$-r and $\forall$-l rule, respectively. P is:

$$
\cfrac{
\cfrac{P_1}{\Delta_1 \Rightarrow \Gamma_1, {}^{L,R}\alpha}{\Delta_1 \Rightarrow \Gamma_1, {}^L\forall R.\alpha} \;\forall\text{-r}
\qquad
\cfrac{\cfrac{P_2}{\Delta_2, {}^{L,R}\alpha \Rightarrow \Gamma_2}}{\Delta_2, {}^L\forall R.\alpha \Rightarrow \Gamma_2} \;\forall\text{-l}
}{\Delta_1, \Delta_2 \Rightarrow \Gamma_1, \Gamma_2} \;({}^L\forall R.\alpha)
$$

which again by hypothesis, none of the proofs P_n for $n \in \{1, 2\}$ contain a mix. These proof can be transformed into:

$$
\cfrac{
\cfrac{P_1}{\Delta_1 \Rightarrow \Gamma_1, {}^{L,R}\alpha}
\qquad
\cfrac{P_2}{\Delta_2, {}^{L,R}\alpha \Rightarrow \Gamma_2}
}{\Delta_1, \Delta_2 \Rightarrow \Gamma_1, \Gamma_2} \;({}^{L,R}\alpha)
$$

which contains one mix with grade less than $g({}^L\forall R.\alpha)$. So by induction hypothesis, we can obtain a proof which contains no mixes.

(iv) The outermost logical symbol is $\exists$. The treatment is similar to the case above.

(v) The outermost logical symbol is $\neg$ and P is:

$$\frac{\begin{array}{c}P_1\\[2pt]\Delta_1, {}^L\alpha \Rightarrow \Gamma_1\end{array}}{\Delta_1 \Rightarrow \Gamma_1, {}^{\neg L}\neg\alpha}\text{ neg-r} \qquad \frac{\begin{array}{c}P_2\\[2pt]\Delta_2 \Rightarrow \Gamma_2, {}^L\alpha\end{array}}{\Delta_2, {}^{\neg L}\neg\alpha \Rightarrow \Gamma_2}\text{ neg-l}$$

$$\frac{}{\Delta_1, \Delta_2 \Rightarrow \Gamma_1, \Gamma_2}\ ({}^{\neg L}\neg\alpha)$$

This proof can be transformed into:

$$\frac{\begin{array}{cc}\dfrac{P_2}{\Delta_2 \Rightarrow \Gamma_2, {}^L\alpha} & \dfrac{P_1}{\Delta_1, {}^L\alpha \Rightarrow \Gamma_1}\end{array}}{\dfrac{\Delta_2, \Delta_1 \Rightarrow \Gamma_2, \Gamma_1}{\Delta_1, \Delta_2 \Rightarrow \Gamma_1, \Gamma_2}\text{ perm*}}\ ({}^L\alpha)$$

which contains one mix with grade less than $g({}^{\neg L}\neg\alpha)$. So by the induction hypothesis, we can obtain a proof which contains no mixes.

(1.9) Both S_1 and S_2 are lower sequents of logical inferences, $rank_l(P) = rank_r(P) = 1$ and J being a *quasi-mix* $(\gamma, {}^{+\exists R}\gamma)$ where the mix formulas on each side are the principal formula of the logical inferences. Let us here present just the case $\sqcup$. In this case S_1 and S_2 must be lower sequents of $\sqcup$-r and $\sqcup$-l rule, respectively:

$$\frac{\dfrac{P_1}{\delta \Rightarrow \Gamma_1, {}^L\alpha, {}^L\beta}\ \sqcup\text{-r}}{\dfrac{\delta \Rightarrow \Gamma_1, {}^L(\alpha \sqcup \beta)}{}} \qquad \frac{\dfrac{P_2}{\Delta_2, {}^{\exists R, L}\alpha \Rightarrow \Gamma_2}\quad \dfrac{P_3}{\Delta_2, {}^{\exists R, L}\beta \Rightarrow \Gamma_2}}{\Delta_2, {}^{\exists R, L}(\alpha \sqcup \beta) \Rightarrow \Gamma_2}\ \sqcup\text{-l}$$

$$\frac{}{{}^{+\exists R}\delta, \Delta_2 \Rightarrow {}^{+\exists R}\Gamma_1, \Gamma_2}\ ({}^L(\alpha \sqcup \beta), {}^{\exists R, L}(\alpha \sqcup \beta))$$

This proof can be transformed into:

$$\frac{\dfrac{\dfrac{P_1}{\delta \Rightarrow \Gamma_1, {}^L\alpha, {}^L\beta}\quad \dfrac{P_2}{\Delta_2, {}^{\exists R, L}\alpha \Rightarrow \Gamma_2}}{{}^{+\exists R}\delta, \Delta_2 \Rightarrow {}^{+\exists R}\Gamma_1, {}^{\exists R, L}\beta, \Gamma_2}\ ({}^L\alpha, {}^{\exists R, L}\alpha)\quad \dfrac{P_3}{\Delta_2, {}^{\exists R, L}\beta \Rightarrow \Gamma_2}\ ({}^{\exists R, L}\beta)}{\dfrac{{}^{+\exists R}\delta, \Delta_2, \Delta_2 \Rightarrow {}^{+\exists R}\Gamma_1, \Gamma_2, \Gamma_2}{{}^{+\exists R}\delta, \Delta_2 \Rightarrow {}^{+\exists R}\Gamma_1, \Gamma_2}\text{ perm*; contract*}}$$

which again contains one *mix* and one *quasi-mix*, but both with grade less than the grade of *quasi-mix* on P. So by the induction hypothesis, we can obtain a proof which contains no *quasi-mixes* at all. All other cases of outermost logical symbol in *quasi-mix* inferences can be obtained in a similar way.

Case 2: $rank > 2$, i.e., $rank_l(P) > 1$ and/or $rank_r(P) > 1$

The induction hypothesis is that from every proof Q which contains a *quasi-mix* only as the last inference, and which satisfies either $g(Q) < g(P)$, or $g(Q) = g(P)$ and $rank(Q) < rank(P)$, we can eliminate the application of the *quasi-mix*.

(2.1) $rank_r(P) > 1$

(2.1.1) Let us consider a *quasi-mix* of the form $\left({}^L\alpha, {}^{\exists R,L}\alpha\right)$ in which Γ_2 contains ${}^{\exists R,L}\alpha$ or ${}^L\delta$ is ${}^L\alpha$. In this case, we construct a new proof as follows.

$$
\cfrac{
\cfrac{
\cfrac{
\cfrac{{}^L\delta \Rightarrow \Gamma_1}{{}^{\exists R,L}\delta \Rightarrow {}^{+\exists R}\Gamma_1}\ \text{prom-}\exists
}{{}^{\exists R,L}\delta \Rightarrow {}^{+\exists R}\Gamma_1^*,\ {}^{\exists R,L}\alpha}\ \text{perm*; contract*}
}{{}^{\exists R,L}\delta,\ \Delta_2^* \Rightarrow {}^{+\exists R}\Gamma_1^*,\ \Gamma_2}\ \text{weak*; perm*}
}{}
$$

where the assumption Γ_2 contains ${}^{\exists R,L}\alpha$ was used in the last inference to construct Γ_2. When Δ_1 is ${}^L\alpha$, we construct a new proof as follows:

$$
\cfrac{
\cfrac{
\Delta_2 \Rightarrow \Gamma_2
}{{}^{\exists R,L}\alpha,\ \Delta_2^* \Rightarrow \Gamma_2}\ \text{perm*; weak*}
}{{}^{\exists R,L}\alpha,\ \Delta_2^* \Rightarrow {}^{+\exists R}\Gamma_1^*,\ \Gamma_2}\ \text{perm*; weak*}
$$

(2.1.2) S_2 is the lower sequent of a inference J_2, where J_2 is not a logical inference whose principal formula is δ. We will consider just the case where the *quasi-mix* is of the form $(\delta, {}^{+\exists R}\delta)$, the other cases of quasi-mix can be treated in a similar way. P has the form:

$$
\cfrac{
\Delta_1 \Rightarrow \Gamma_1 \qquad
\cfrac{
\begin{matrix} P_2 \\ \Phi \Rightarrow \Psi \end{matrix} \qquad \Delta_2 \Rightarrow \Gamma_2
}{}\ J_2
}{\Delta_1,\ \Delta_2^* \Rightarrow \Gamma_1^*,\ \Gamma_2}\ (\delta, {}^{+\exists R}\delta)
$$

where P_1 and P_2 contain no *quasi-mixes* and Φ contains at least one occurrence of ${}^{+\exists R}\delta$. We first consider the proof P':

$$
\cfrac{
\begin{matrix} P_1 \\ \Delta_1 \Rightarrow \Gamma_1 \end{matrix} \qquad
\begin{matrix} P_2 \\ \Phi \Rightarrow \Psi \end{matrix}
}{\Delta_1,\ \Phi^* \Rightarrow \Gamma_1^*,\ \Psi}\ (\delta, {}^{+\exists R}\delta)
$$

$g(P) = g(P')$, $rank_l(P') = rank_l(P)$ and $rank_r(P') = rank_r(P) - 1$. Thus, by the induction hypothesis, the final sequent in P' is provable without *quasi-mix*. Given that, we can now construct a proof:

$$
\cfrac{
\cfrac{
\begin{matrix} P' \\ \Delta_1,\ \Phi^* \Rightarrow \Gamma_1^*,\ \Psi \end{matrix}
}{\Phi^*,\ \Delta_1 \Rightarrow \Gamma_1^*,\ \Psi}\ \text{perm*}
}{\Delta_2^*,\ \Delta_1 \Rightarrow \Gamma_1^*,\ \Gamma_2}\ J_2
$$

In the case that the auxiliary formula in J_2 in P is a mix in Φ, we need an additional weakening before J_2 in the last proof.

(2.1.3) Δ_1 contains no δ's, S_2 is the lower sequent of a logical inference whose principal formula is δ and J is a *mix* rule inference. We have to consider several cases according to the outermost logical symbol of δ:

(i) The outermost logical symbol of δ is $\sqcap$. The last part of P is of the form:

$$J \cfrac{P_1 \qquad \cfrac{P_2}{\Delta_2, {}^L\alpha, {}^L\beta \Rightarrow \Gamma_2}{\Delta_2, {}^L(\alpha \sqcap \beta) \Rightarrow \Gamma_2} J_2 \; \left({}^L{(\alpha \sqcap \beta)}\right)}{\Delta_1, \Delta_2^* \Rightarrow \Gamma_1^*, \Gamma_2}$$

Now let us consider the proof Q:

$$J \cfrac{P_1 \qquad \cfrac{P_2}{\Delta_2, {}^L\alpha, {}^L\beta \Rightarrow \Gamma_2} \; \left({}^L{(\alpha \sqcap \beta)}\right)}{\Delta_1, \Delta_2^*, {}^L\alpha, {}^L\beta \Rightarrow \Gamma_1^*, \Gamma_2}$$

assuming that ${}^L(\alpha \sqcap \beta)$ is in Δ_2. Note that $g(Q) = g(P)$, $rank_l(Q) = rank_l(P)$ and $rank_r(Q) = rank_r(P) - 1$. Hence by the induction hypothesis, the end-sequent of Q is provable without a mix. Let us call such proof Q' and consider the following proof P':

$$J \cfrac{P_1 \qquad \cfrac{\cfrac{Q'}{\Delta_1, \Delta_2^*, {}^L\alpha, {}^L\beta \Rightarrow \Gamma_1^*, \Gamma_2}}{\Delta_1, \Delta_2^*, {}^L(\alpha \sqcap \beta) \Rightarrow \Gamma_1^*, \Gamma_2} J_2 \; \left({}^L{(\alpha \sqcap \beta)}\right)}{\Delta_1, \Delta_1, \Delta_2^* \Rightarrow \Gamma_1^*, \Gamma_1^*, \Gamma_2}$$

Given that, $g(P') = g(P)$, $rank_l(P') = rank_l(P)$ and $rank_r(P') = 1$ (for Δ_1 contains no occurences of ${}^L(\alpha \sqcap \beta)$) by the induction hypothesis the end-sequent of P' is provable without a mix, and so is the end-sequent of P.

(ii) The outermost logical symbol of δ is $\sqcup$. Let us consider a proof P whose last part is of the form:

$$\cfrac{P_1 \qquad \cfrac{\cfrac{P_2}{\Delta_2, {}^L\alpha \Rightarrow \Gamma_2} \qquad \cfrac{P_3}{\Delta_2, {}^L\beta \Rightarrow \Gamma_2}}{\Delta_2, {}^L(\alpha \sqcup \beta) \Rightarrow \Gamma_2} \; ({}^L(\alpha \sqcup \beta))}{\Delta_1, \Delta_2^* \Rightarrow \Gamma_1^*, \Gamma_2}$$

Assuming that ${}^L(\alpha \sqcup \beta)$ is in P_1 and P_2, consider the proof Q_1:

$$\cfrac{\cfrac{P_1}{\Delta_1 \Rightarrow \Gamma_1} \qquad \cfrac{P_2}{\Delta_2, {}^L\alpha \Rightarrow \Gamma_2}}{\Delta_1, \Delta_2^*, {}^L\alpha \Rightarrow \Gamma_1^*, \Gamma_2} \; ({}^L(\alpha \sqcup \beta))$$

and Q_2:

$$\frac{P_1 \qquad\qquad P_3}{\Delta_1 \Rightarrow \Gamma_1 \qquad \Delta_2, {}^L\beta \Rightarrow \Gamma_2}{\Delta_1, \Delta_2^*, {}^L\beta \Rightarrow \Gamma_1^*, \Gamma_2} \; ({}^L{(\alpha \sqcup \beta)})$$

We note that $g(Q_1) = g(Q_2) = g(P)$, $rank_l(Q_1) = rank_l(Q_2) = rank_l(P)$ and $rank_r(Q_1) = rank_r(Q_2) < rank_r(P)$. Hence, by the induction hypothesis, the end-sequents of P_1 and P_2 are provable without a *mix*. Let us consider new proofs without *mix* Q_1' and Q_2' in the construction of P':

$$\frac{P_1 \qquad \dfrac{Q_1' \qquad\qquad Q_2'}{\dfrac{\Delta_1, \Delta_2^*, {}^L\alpha \Rightarrow \Gamma_1^*, \Gamma_2 \qquad \Delta_1, \Delta_2^*, {}^L\beta \Rightarrow \Gamma_1^*, \Gamma_2}{\Delta_1, \Delta_2^*, {}^L(\alpha \sqcup \beta) \Rightarrow \Gamma_1^*, \Gamma_2}\;\sqcup\text{-}l}}{\Delta_1, \Delta_1, \Delta_2^* \Rightarrow \Gamma_1^*, \Gamma_1^*, \Gamma_2}\;({}^L{(\alpha \sqcup \beta)})$$

Then, $g(P') = g(P)$, $rank_l(P') = rank_l(P)$ and $rank_r(P') = 1$, since Δ_1 and Δ_2^* do not contain ${}^L(\alpha \sqcup \beta)$. By the induction hypothesis the end-sequent of P' is provable without a *mix*.

(iii) The outermost logical symbol of δ is $\forall$. That is, the mix formula is of the form ${}^L(\forall R.\alpha)$. Let us consider the proof P:

$$\frac{P_1 \qquad \dfrac{P_2}{\dfrac{\Delta_2, {}^{L,R}\alpha \Rightarrow \Gamma_2}{\Delta_2, {}^L(\forall R.\alpha) \Rightarrow \Gamma_2}\;\forall\text{-}l}}{\Delta_1 \Rightarrow \Gamma_1 \qquad}{\Delta_1, \Delta_2^* \Rightarrow \Gamma_1^*, \Gamma_2}\;\left({}^L\forall R.\alpha\right)$$

where ${}^L\forall R.\alpha$ occurs on Δ_2 since $rank_r(P) > 1$. Let us consider a proof Q as follows:

$$\frac{P_1 \qquad\qquad P_2}{\Delta_1 \Rightarrow \Gamma_1 \qquad \Delta_2, {}^{L,R}\alpha \Rightarrow \Gamma_2}{\Delta_1, \Delta_2^*, {}^{L,R}\alpha \Rightarrow \Gamma_1^*, \Gamma_2}\;\left({}^L\forall R.\alpha\right)$$

Note that $grade(Q) = grade(P)$, $rank_l(Q) = rank_l(P)$ and $rank_r(Q) = rank_r(P) - 1$. So, by the induction hypothesis one can obtain a proof Q' with the same end-sequent as Q without *quasi-mix* inferences. Now consider the new proof P':

$$\frac{P_1 \qquad \dfrac{Q'}{\dfrac{\Delta_1, \Delta_2^*, {}^{L,R}\alpha \Rightarrow \Gamma_1^*, \Gamma_2}{\Delta_1, \Delta_2^*, {}^L\forall R.\alpha \Rightarrow \Gamma_1^*, \Gamma_2}\;\forall\text{-}l}}{\dfrac{\Delta_1, \Delta_1, \Delta_2^* \Rightarrow \Gamma_1^*, \Gamma_1^*, \Gamma_2}{\Delta_1, \Delta_2^* \Rightarrow \Gamma_1^*, \Gamma_2}\;\text{contract}^*}\;\left({}^L\forall R.\alpha\right)$$

Now we have

$$g(P') = g(P) \quad \text{and} \quad rank_l(P') = rank_l(P)$$

and $rank_r(P') = 1$ (for Δ_1 and Δ_2^* do not contain $^L\forall R.\alpha$). Thus the end-sequent of P' (the same of P) is provable without *quasi-mix* by the induction hypothesis.

The remaining cases where δ is of the form $^L\exists R.\alpha$ and $^L\neg\alpha$ are treated in a similar way.

(2.1.4)   The same conditions that hold for 2.1.3 but J is a *quasi-mix* rule inference. We have to consider several cases according to the outermost logical symbol of δ. All the cases are treated in a similar way of cases 2.1.3.

(2.2) $rank_r(P) = 1$ and $rank_l(P) > 1$. This case is proved as in case 2.1 above.

$\square$

References

1. Borgida, A., Franconi, E., Horrocks, I., McGuinness, D.L., Patel-Schneider, P.F.: Explaining ALC subsumption. In: Lambrix, P., Borgida, A., Lenzerini, M., Möller, R., Patel-Schneider, P.F. (eds.) Proceedings of the 1999 International Workshop on Description Logics 1999, vol. 22. Linköping, Sweden (1999). http://SunSITE.Informatik.RWTH-Aachen.DE/Publications/CEUR-WS/Vol-22/borgida.ps
2. Buss, S.R.: An introduction to proof theory. In: Buss, S.R. (ed.) Handbook of Proof Theory, Studies in Logic and the Foundations of Mathematics, p. 811. Elsevier, Amsterdam (1998)
3. Takeuti, G.: Proof Theory. Number 81 in Studies in Logic and the Foundations of Mathematics. North-Holland, Amsterdam (1975)

Chapter 4
Comparing SC$_{\mathcal{ALC}}$ With Other $\mathcal{ALC}$ Deduction Systems

Abstract The structural subsumption algorithm is restricted to a quite inexpressive language. Simple Tableaux based algorithms generally fails to provide short proofs. On the other hand, the latter has a useful property, it returns a counter-model from an unsuccessful proof. A counter-model, that is, an interpretation that falsifies the premise, is a quite useful object to a knowledge-base engineer. In this chapter we compare our SC$_{\mathcal{ALC}}$ system with the structural subsumption algorithm and the Tableaux algorithm for $\mathcal{ALC}$.

Keywords Structure subsumption algorithm · Tableaux for $\mathcal{ALC}$ · Sequent calculus · Proof explanation · Counter example · Counter model

4.1 Introduction

In Sect. 4.2 we compare SC$_{\mathcal{ALC}}$ with the structural subsumption algorithm. In Sect. 4.3 we show how to extend SC$_{\mathcal{ALC}}$ in order to be able to construct a counter-model from unsuccessful proofs. In this way, SC$_{\mathcal{ALC}}$ can be compared with Tableaux algorithms, indeed. In fact the system that will be presented in the section, SC$^{[]}{}_{\mathcal{ALC}}$, is a structural-free sequent calculus designed to provide sequent proofs without considering backtracking during the proof-construction from conclusion to axioms. Nevertheless, two secondary results are obtained from Sect. 4.3:

1. In Sect. 3.3 a relative completeness of SC$_{\mathcal{ALC}}$ regarding the axiomatic presentation of $\mathcal{ALC}$ is shown. In Sect. 4.3 we present an alternative proof of SC$_{\mathcal{ALC}}$ completeness. The method used in this section is a basis for constructing a proof of SC$_{\mathcal{ALCQI}}$ completeness.
2. Since the results of Sect. 4.3 are obtained from a SC$_{\mathcal{ALC}}$ without cut-rule, we are actually proving the completeness of SC$_{\mathcal{ALC}}$ without the cut-rule. Given that, the results can also be considered an alternative method of cut-elimination for

A. Rademaker, *A Proof Theory for Description Logics*,
SpringerBriefs in Computer Science, DOI: 10.1007/978-1-4471-4002-3_4,
© The Author(s) 2012

the SC$_{\mathcal{ALC}}$ presented in Sect. 3.4, where we followed Gentzen's original proof for cut elimination.

4.2 Comparing $\mathcal{SALC}$ With the Structural Subsumption Algorithm

The *structural subsumption algorithms* (**SSA**) presented in [1] compare the syntactic structure of two normalized concept descriptions in order to verify if the first one is subsumed by the second one. In order to compare deductions in SC$_{\mathcal{ALC}}$ with deductions in **SSA**, we just have to observe that each step taken by a bottom-up construction of a SC$_{\mathcal{ALC}}$ proof corresponds to a step of the **SSA** algorithm towards the concepts matching. Moreover, **SSA** can deal with concepts expressed in $\mathcal{ALN}$ language ($\mathcal{AL}$ augmented with number restrictions). On the other hand, SC$_{\mathcal{ALC}}$ can deal with concepts expressed in $\mathcal{ALC}$ and will be extended in Chap. 6 to deal with $\mathcal{ALCQI}$.

For a concrete example, let us consider the SC$_{\mathcal{ALC}}$ proof below where A and B stand for atomic concepts and C and D for normalized concepts.

$$
\cfrac{
\cfrac{A_1 \Rightarrow B_1}{
\cfrac{\forall R_1.C_1, A_1 \Rightarrow B_1}{A_1, \forall R_1.C_1 \Rightarrow B_1}}
\qquad
\cfrac{
\cfrac{
\cfrac{^{R_1}C_1 \Rightarrow {}^{S_1}D_1}{^{R_1}C_1 \Rightarrow \forall S_1.D_1}}{\forall R_1.C_1 \Rightarrow \forall S_1.D_1}}{A_1, \forall R_1.C_1 \Rightarrow \forall S_1.D_1}
}{
\cfrac{A_1, \forall R_1.C_1 \Rightarrow B_1 \sqcap \forall S_1.D_1}{A_1 \sqcap \forall R_1.C_1 \Rightarrow B_1 \sqcap \forall S_1.D_1}}
$$

The deduction above deals with two normalized concepts, $A_1 \sqcap \forall R_1.C_1$ and $B_1 \sqcap \forall S_1.D_1$. It would conclude the subsumption (sequent) whenever the top-sequents also ensure their respective subsumptions. This is just what the recursive procedure of **SSA** does.

4.3 Obtaining Counter-Models From Unsuccessful Proof Trees

The SC$_{\mathcal{ALC}}$ system rules are not *deterministic*. That is, if rules are applied in the wrong order, we fail to obtain a proof of an $\mathcal{ALC}$ theorem. For instance, consider the fully expanded proof tree presented in the Example 2. The initial sequent denotes a subsumption proved valid by the Example 1 (page xx). Despite that, reading bottom-up, from the sequent

$$^{\exists child}\top, {}^{\forall child}\neg(\exists child.\neg Doctor) \Rightarrow {}^{\exists child, \forall child}Doctor$$

the rule weak-l was applied to allow the application of rule prom-$\exists$. But the weak-l rule removed the wrong concept from the sequent, which turned the proof impossible to be finished, that is, the top sequent is not an axiom. Given that, in order to obtain a counter-model from unsuccessful proofs, we must consider not only one of the possible fully expanded proof trees but all of them. In other words, one possible fully expanded proof tree of a given sequent is not a sufficient evidence that this sequent is not a theorem.

Example 2 An unsuccessful proof of a valid sequent in $SC_{\mathcal{ALC}}$:

$$
\cfrac{
\cfrac{
\cfrac{
\cfrac{
\cfrac{
\cfrac{
\cfrac{
\cfrac{\top \Rightarrow {}^{\forall child} Doctor}{\exists child \top \Rightarrow {}^{\exists child, \forall child} Doctor} \text{ prom-}\exists
}{\exists child \top, {}^{\forall child} \neg (\exists child.\neg Doctor) \Rightarrow {}^{\exists child, \forall child} Doctor} \text{ weak-l}
}{\exists child \top child, {}^{\forall child} \neg (\exists child.\neg Doctor) \Rightarrow {}^{\exists child} \forall child.Doctor} \text{ }\forall\text{-r}
}{\exists child \top, {}^{\forall child} \neg (\exists child.\neg Doctor) \Rightarrow \exists child.\forall child.Doctor} \text{ }\exists\text{-r}
}{\exists child \top, \forall child.\neg (\exists child.\neg Doctor) \Rightarrow \exists child.\forall child.Doctor} \text{ }\forall\text{-l}
}{\exists child.\top, \forall child.\neg (\exists child.\neg Doctor) \Rightarrow \exists child.\forall child.Doctor} \text{ }\exists\text{-l}
}{\exists child.\top \sqcap \forall child.\neg (\exists child.\neg Doctor) \Rightarrow \exists child.\forall child.Doctor} \text{ }\sqcap\text{-l}
$$

To better illustrate the problem with obtaining a counter-model from fully expanded proof trees, consider Example 3 (page xx) which does not hold for concepts A and B in general.

Example 3 Two possibly fully expanded proof trees for the invalid subsumption:

$$\exists R.A \sqcap \exists R.B \sqsubseteq \exists R.(A \sqcap B)$$

$$
\cfrac{
\cfrac{
\cfrac{
\cfrac{
\cfrac{
\cfrac{
\cfrac{B \Rightarrow A \qquad B \Rightarrow B}{B \Rightarrow A \sqcap B} \text{ }\sqcap\text{-r}
}{{}^{\exists R}B \Rightarrow {}^{\exists R}A \sqcap B} \text{ prom-}\exists
}{{}^{\exists R}A, {}^{\exists R}B \Rightarrow {}^{\exists R}A \sqcap B} \text{ weak-l}
}{{}^{\exists R}A, {}^{\exists R}B \Rightarrow \exists R.A \sqcap B} \text{ }\exists\text{-r}
}{{}^{\exists R}A, \exists R.B \Rightarrow \exists R.A \sqcap B} \text{ }\exists\text{-l}
}{\exists R.A, \exists R.B \Rightarrow \exists R.A \sqcap B} \text{ }\exists\text{-l}
}{\exists R.A \sqcap \exists R.B \Rightarrow \exists R.A \sqcap B} \text{ }\sqcap\text{-l}
\qquad\qquad
\cfrac{
\cfrac{
\cfrac{
\cfrac{
\cfrac{
\cfrac{
\cfrac{A \Rightarrow A \qquad A \Rightarrow B}{A \Rightarrow A \sqcap B} \text{ }\sqcap\text{-r}
}{{}^{\exists R}A \Rightarrow {}^{\exists R}A \sqcap B} \text{ prom-}\exists
}{{}^{\exists R}A, {}^{\exists R}B \Rightarrow {}^{\exists R}A \sqcap B} \text{ weak-l}
}{{}^{\exists R}A, {}^{\exists R}B \Rightarrow \exists R.A \sqcap B} \text{ }\exists\text{-r}
}{{}^{\exists R}A, \exists R.B \Rightarrow \exists R.A \sqcap B} \text{ }\exists\text{-l}
}{\exists R.A, \exists R.B \Rightarrow \exists R.A \sqcap B} \text{ }\exists\text{-l}
}{\exists R.A \sqcap \exists R.B \Rightarrow \exists R.A \sqcap B} \text{ }\sqcap\text{-l}
$$

Given a fully expanded proof tree, in the attempt to construct a counter-model for the bottom sequent, the process must start from the most top sequents, not axioms, going into the direction of the bottom sequent adjusting the model at each rule application in order to guarantee that in each step, if the model being constructed does not satisfy the premise, it should not satisfy the conclusion. In this way we would have an algorithm to construct a counter-model for any fully expanded proof tree.

In Example 3, let us first consider the fully expanded proof tree on the left, if we start from the logical axiom $B \sqsubseteq B$ it will not be possible to construct any counter-model for it. But starting from $B \Rightarrow A$ we can easily construct an interpretation $\mathcal{I}$

where $B^{\mathcal{I}} \not\subseteq A^{\mathcal{I}}$. But this is not sufficient to negate the bottom sequent. Basically, from top-down, when we get into the point to consider the application of rule weak-l, we must note that the formula introduced on the left force us to include one more restriction in the counter-model $\mathcal{I}$ being constructed. $\mathcal{I}$ not only has to guarantee $B^{\mathcal{I}} \not\subseteq A^{\mathcal{I}}$ but also $A^{\mathcal{I}} \not\subseteq B^{\mathcal{I}}$. The derivation on the right would let us conclude these same restrictions in the inverse order. The two derivations are basically the two possibles choices of formulas in the application of weak-l rule.

One important property of weak and promotional rules is that they are not double-sound. A rule is said double-sound if it is not only truth-preserving from the premises to conclusion but also from the conclusion to its premises. Regarding the top-bottom construction of counter-model, this means that in the adjustment of the counter-model $\mathcal{I}$ being constructed, the fact that $\mathcal{I}$ does not satisfy the premise of a weak rule application does not guarantee that it does not satisfy its conclusion. Moreover, the introduced formula by the weak rule can be arbitrary complex making the adjustment of the counter-model not trivial nor modular.

Let us consider the system SC$^{[]}_{\mathcal{ALC}}$, a conservative extension of SC$_{\mathcal{ALC}}$ presented in Fig. 4.1. SC$^{[]}_{\mathcal{ALC}}$ sequents are expressions of the form $\Delta \Rightarrow \Gamma$ where Δ and Γ are *sets* of labeled concepts (possibly frozen). A frozen concept α is represented as $[\alpha]^n$ where n is the index (context identifier) of the frozen concept. The notation $[\Delta]^n$ means that each $\delta \in \Delta$ is frozen with the same index (i.e. $\{[\delta]^n \mid \delta \in \Delta\}$). Given a SC$^{[]}_{\mathcal{ALC}}$ sequent with the general form 4.1, we call each pair (Δ_k, Γ_k) a *context* in the sequent.

$$\Delta_1, [\Delta_2]^1, \ldots, [\Delta_n]^{n-1} \Rightarrow \Gamma_1, [\Gamma_2]^1, \ldots, [\Gamma_n]^{n-1} \tag{4.1}$$

SC$^{[]}_{\mathcal{ALC}}$ does not have permutation, contraction or the *cut* rule from SC$_{\mathcal{ALC}}$. Reading bottom-up, the weak rules of SC$^{[]}_{\mathcal{ALC}}$ *save* the context of the proof before removing a concept from the lefthand (antecedent) or righthand side (succedent) of the sequent and the frozen-exchange changes the contexts during a proof construction. Considering that in SC$^{[]}_{\mathcal{ALC}}$ the sequents are constructed by two sets (not lists) of concepts, weak rules are still necessary only to allow the application of promotional rules.

The notation $^{+\forall R}\Gamma$ or $^{+\exists R}\Gamma$ denotes the addition of the Role R existentialy or universaly quantified in the front of each list of labels of all formulas of Γ. In rules ⊓-{l,r}, ⊔-{l,r}, ∀-{l,r}, ∃-{l,r} and in the axiom, Δ and Γ stand for labeled concepts frozen or not. In the promotional, frozen-exchange and weak rules we have to distinguish the frozen concepts from the non-frozen ones. We use the notation $[\Delta]$ (resp. $[\Gamma]$) to denote the set of all frozen concepts in the sequent regardless their index. The index k must be in all rules a fresh one.

In rule frozen-exchange, all formulas in Δ_2 and Γ_2 cannot be the conclusion of any rule application except the frozen-exchange. This proviso is not actually necessary to guarantee the soundness of the system, it is more a strategy for proof constructions. The idea is to postpone the exchange of contexts until no other rule can reduce the current active context, avoiding unnecessary swapping of contexts.

In Sect. 3.1, we presented the natural interpretation of a sequent $\Delta \Rightarrow \Gamma$ in SC$_{\mathcal{ALC}}$ as the $\mathcal{ALC}$ formula

$$\overline{\Delta, \delta \Rightarrow \delta, \Gamma}$$

$$\frac{[\Delta, \delta]^k, \Delta \Rightarrow \Gamma, [\Gamma]^k}{\Delta, \delta \Rightarrow \Gamma} \text{ weak-l} \qquad \frac{[\Delta]^k, \Delta \Rightarrow \Gamma, [\Gamma, \gamma]^k}{\Delta \Rightarrow \Gamma, \gamma} \text{ weak-r}$$

$$\frac{\Delta, {}^{L, \forall R}\alpha \Rightarrow \Gamma}{\Delta, {}^{L}(\forall R.\alpha)L_2 \Rightarrow \Gamma} \forall\text{-l} \qquad \frac{\Delta \Rightarrow \Gamma, {}^{L, \forall R}\alpha}{\Delta \Rightarrow \Gamma, {}^{L}(\forall R.\alpha)} \forall\text{-r}$$

$$\frac{\Delta, {}^{L, \exists R}\alpha \Rightarrow \Gamma}{\Delta, {}^{L}(\exists R.\alpha) \Rightarrow \Gamma} \exists\text{-l} \qquad \frac{\Delta \Rightarrow \Gamma, {}^{L, \exists R}\alpha}{\Delta \Rightarrow \Gamma, {}^{L}(\exists R.\alpha)} \exists\text{-r}$$

$$\frac{\Delta, {}^{\forall L}\alpha, {}^{\forall L}\beta \Rightarrow \Gamma}{\Delta, {}^{\forall L}(\alpha \sqcap \beta) \Rightarrow \Gamma} \sqcap\text{-l} \qquad \frac{\Delta \Rightarrow \Gamma, {}^{\forall L}\alpha \quad \Delta \Rightarrow \Gamma, {}^{\forall L}\beta}{\Delta \Rightarrow \Gamma, {}^{\forall L}(\alpha \sqcap \beta)} \sqcap\text{-r}$$

$$\frac{\Delta, {}^{\exists L}\alpha \Rightarrow \Gamma \quad \Delta, {}^{\exists L}\beta \Rightarrow \Gamma}{\Delta, {}^{\exists L}(\alpha \sqcup \beta) \Rightarrow \Gamma} \sqcup\text{-l} \qquad \frac{\Delta \Rightarrow \Gamma, {}^{\exists L}\alpha, {}^{\exists L}\beta}{\Delta \Rightarrow \Gamma, {}^{\exists L}(\alpha \sqcup \beta)} \sqcup\text{-r}$$

$$\frac{\Delta \Rightarrow \Gamma, {}^{\neg L}\alpha}{\Delta, {}^{L}\neg\alpha \Rightarrow \Gamma} \neg\text{-l} \qquad \frac{\Delta, {}^{\neg L}\alpha \Rightarrow \Gamma}{\Delta \Rightarrow \Gamma, {}^{L}\neg\alpha} \neg\text{-r}$$

$$\frac{[\Delta], {}^{L}\delta \Rightarrow \Gamma, [\Gamma_1]}{[\Delta], {}^{\exists R, L}\delta \Rightarrow {}^{+\exists R}\Gamma, [\Gamma_1]} \text{ prom-}\exists \qquad \frac{[\Delta_1], \Delta \Rightarrow {}^{L}\gamma, [\Gamma]}{[\Delta_1], {}^{+\forall R}\Delta \Rightarrow {}^{\forall R, L}\gamma, [\Gamma]} \text{ prom-}\forall$$

$$\frac{[\Delta], [\Delta_2]^k, \Delta_1 \Rightarrow \Gamma_1, [\Gamma_2]^k, [\Gamma]}{[\Delta], \Delta_2, [\Delta_1]^n \Rightarrow [\Gamma_1]^n, \Gamma_2, [\Gamma]} \text{ frozen-exchange}$$

Fig. 4.1 The System $SC^{\sqcap}{}_{\mathcal{ALC}}$

$$\prod_{\delta \in \Delta} \sigma(\delta) \sqsubseteq \bigsqcup_{\gamma \in \Gamma} \sigma(\gamma)$$

Given an interpretation function $.^{\mathcal{I}}$ we write $\mathcal{I} \models \Delta \Rightarrow \Gamma$, if and only if,

$$\bigcap_{\delta \in \Delta} \sigma(\delta)^{\mathcal{I}} \subseteq \bigcup_{\gamma \in \Gamma} \sigma(\gamma)^{\mathcal{I}}$$

Now we have to extend that definition to give the *semantics* of the sequents with (indexed) frozen concepts.

Definition 12 (Satisfiability of frozen-labeled sequents) Let $\Delta \Rightarrow \Gamma$ be a sequent with its succedent and antecedent having formulas that range over labeled concepts and frozen labeled concepts. This sequent has the general form of 4.1. Let

$(\mathcal{I}_1, \ldots, \mathcal{I}_n)$ be a tuple of interpretations. We say that this tuple satisfies $\Delta \Rightarrow \Gamma$ if and only if, one of the following clauses holds:

$$\mathcal{I}_1 \models \Delta_1 \Rightarrow \Gamma_1 \quad \mathcal{I}_2 \models \Delta_2 \Rightarrow \Gamma_2 \quad \ldots \quad \mathcal{I}_n \models \Delta_n \Rightarrow \Gamma_n \qquad (4.2)$$

That is, the first projection should satisfy the set of non-frozen formulas. The second projection should satisfy the set of frozen formulas with the minimum index and so on. The sequent $\Delta \Rightarrow \Gamma$ is not satisfiable by a tuple of interpretations, if and only if, no interpretation in the tuple satisfy its corresponding *context*.

Before proceeding to present the procedure to obtain counter-models from SC$^{[]}{}_{\mathcal{ALC}}$-proofs, we must introduce Lemma 5 showing that SC$^{[]}{}_{\mathcal{ALC}}$ is a conservative extension of SC$_{\mathcal{ALC}}$.

Lemma 5 *Consider $\Delta \Rightarrow \Gamma$ a SC$_{\mathcal{ALC}}$ sequent. If P is a proof of $\Delta \Rightarrow \Gamma$ in SC$^{[]}{}_{\mathcal{ALC}}$ then it is possible to construct a proof P' of $\Delta \Rightarrow \Gamma$ in SC$_{\mathcal{ALC}}$.*

Proof Each application of a frozen-exchange rule correspond to a shift of contexts during the bottom-up proof construction process. To proof Lemma 5 we need a two-steps procedure to: (1) remove all frozen-exchange applications of a given proof (a proof in SC$^{[]}{}_{\mathcal{ALC}}$ without any frozen-exchange application is naturally translated to a proof in SC$_{\mathcal{ALC}}$); (2) replace the weak rules of SC$^{[]}{}_{\mathcal{ALC}}$ by their counterparts in SC$_{\mathcal{ALC}}$.

We show that a proof P in SC$^{[]}{}_{\mathcal{ALC}}$ can always be transformed into a proof P' in SC$^{[]}{}_{\mathcal{ALC}}$ without any frozen-exchange rule applications by induction over the number of applications of frozen-exchange occurring in a proof P. Let us consider a topmost application of rule frozen-exchange in P, where reading bottom-up, the frozen-exchange rule recovers a context that was frozen by the γ rule that can be a weak-l or weak-r rule.

$$\Delta, \alpha \Rightarrow \alpha, \Gamma$$
$$\Pi_1$$
$$\frac{[\Delta], [\Delta_1'']^j, \Delta_1 \Rightarrow \Gamma_1, [\Gamma_1'']^j, [\Gamma]}{[\Delta], [\Delta_1]^k \Delta_1'' \Rightarrow \Gamma_1'', [\Gamma_1,]^k, [\Gamma]} \text{ frozen-exchange}$$
$$\Pi_2$$
$$\frac{[\Delta], [\Delta_1]^k \Delta_1' \Rightarrow \Gamma_1', [\Gamma_1,]^k, [\Gamma]}{[\Delta], \Delta_1 \Rightarrow \Gamma_1, [\Gamma]} \gamma$$

We can obtain P' below by simple discarding the proof fragment Π_2.

$$\Delta, \alpha \Rightarrow \alpha, \Gamma$$
$$\Pi_1$$
$$[\Delta], \Delta_1 \Rightarrow \Gamma_1, [\Gamma]$$

Applying recursively the transformations above from top to bottom we obtain a proof in SC$^{[]}{}_{\mathcal{ALC}}$ without any frozen-exchange rule application. Note also that this procedure will remove any *branch* created between the rule that introduced the frozen-formulas and the removed frozen-exchange application.

Given a frozen-exchange free $SC^{[]}_{\mathcal{ALC}}$-proof, to obtain a $SC_{\mathcal{ALC}}$-proof, we only have to drop out the frozen concepts and substitute weak-{l,r} rules application of $SC^{[]}_{\mathcal{ALC}}$ for their counter-parts in $SC_{\mathcal{ALC}}$.

Let us consider the weak-l case, rule weak-r can be dealt similarly. Given the $SC^{[]}_{\mathcal{ALC}}$-proof fragment below containing the top most application of rule weak-r:

$$\frac{\begin{array}{c} \Pi \\ [\Delta], [\Delta_1]^k \Rightarrow \Gamma_2, [\gamma, \Gamma_2]^k, [\Gamma] \end{array}}{[\Delta], \Delta_1 \Rightarrow \gamma, \Gamma_2, [\Gamma]} \text{ weak-r}$$

From the fragment above, we construct:

$$\frac{\begin{array}{c} \Pi \\ \Delta_1 \Rightarrow \Gamma_2 \end{array}}{\Delta_1 \Rightarrow \gamma, \Gamma_2} \text{ weak-r}$$

Applying recursively the transformations above from top to bottom we obtain a proof in $SC_{\mathcal{ALC}}$ from a proof in $SC^{[]}_{\mathcal{ALC}}$. $\qquad\square$

Let us give a precise definition of *fully expanded proof tree*. A fully expanded proof tree of $\Delta \Rightarrow \Gamma$ is a tree having $\Delta \Rightarrow \Gamma$ as root, each internal node being a sequent premise of a valid $SC^{[]}_{\mathcal{ALC}}$ rule application, and each leaf being either a $SC^{[]}_{\mathcal{ALC}}$ axiom (initial sequent) or a top-sequent (not axiom) with not necessarily only atomic concepts. A sequent is a top-sequent if and only if it does not contain *reducible contexts*. A reducible context is a context that if active could be further reduced. In the following lemmas we are interested in fully expanded proof trees that are not $SC^{[]}_{\mathcal{ALC}}$ proofs.

If we consider a particular strategy of rule applications, any fully expanded proof tree will have a special form called *normal form*. The following are the main properties of this strategy:

1. It is a fair strategy of rules applications that avoid infinite loops of, for instance, frozen-exchange applications swapping contexts or unnecessary repetition of proof fragments;
2. Promotional rules will be applied whenever possible, that is, they have high priority over the other rules;
3. The strategy will discard contexts created by the successive application of weak rules and avoid further applications of weak rules once it is possible to detected that they will not be useful to obtain an initial sequent. For instance, from a sequent $\Delta \Rightarrow \Gamma$, where Δ and Γ only contain atomic concept names without any common concept name, we know that using weak rules we would not obtain an initial sequent. Moreover, weak rules will be used with the unique purpose of enabling promotion rules applications.

A more insightful definition of the last item above would be possible if we replace the weak rules in SC$^{[]}_{\mathcal{ALC}}$ by the *weak** rule below.

$$\frac{[\Delta'], [\Delta, \Delta_1]^k, \Delta \Rightarrow \Gamma, [\Gamma_1, \Gamma]^k, [\Gamma']}{[\Delta'], \Delta, \Delta_1 \Rightarrow \Gamma_1, \Gamma, [\Gamma']} \; weak^*$$

Lemma 6 *The weak* rule is a derived rule in* SC$^{[]}_{\mathcal{ALC}}$.

Proof To prove Lemma 6, given a derivation fragment Π with a *weak** rule application, we show how to replace it by successive weak-l and weak-r applications. Without lost of generality, let us consider one special case of *weak** freezing two concepts of both sides of a sequent.

$$\Pi'$$
$$\frac{[\Delta, \delta_1, \delta_2]^k, \Delta \Rightarrow \Gamma, [\gamma_1, \gamma_2, \Gamma]^k}{\Delta, \delta_1, \delta_2 \Rightarrow \gamma_1, \gamma_2, \Gamma} \; weak^*$$

The corresponding fragment Π_1 is presented below. The context k is now followed by the contexts $k+1, k+2, k+3$.

$$\Pi'$$
$$\frac{\frac{\frac{\frac{[\Delta, \delta_1, \delta_2]^k, [\Delta, \delta_2]^{k+1}, [\Delta]^{k+2}, [\Delta]^{k+3}, \Delta \Rightarrow \Gamma, [\gamma_1, \gamma_2, \Gamma]^k, [\gamma_1, \gamma_2, \Gamma]^{k+1}, [\gamma_1, \gamma_2, \Gamma]^{k+2}, [\gamma_2, \Gamma]^{k+3}}{[\Delta, \delta_1, \delta_2]^k, [\Delta, \delta_2]^{k+1}, [\Delta]^{k+2}, \Delta \Rightarrow \gamma_2, \Gamma, [\gamma_1, \gamma_2, \Gamma]^k, [\gamma_1, \gamma_2, \Gamma]^{k+1}, [\gamma_1, \gamma_2, \Gamma]^{k+2}} \; \text{weak-r}}{[\Delta, \delta_1, \delta_2]^k, [\Delta, \delta_2]^{k+1}, \Delta \Rightarrow \gamma_1, \gamma_2, \Gamma, [\gamma_1, \gamma_2, \Gamma]^k, [\gamma_1, \gamma_2, \Gamma]^{k+1}} \; \text{weak-r}}{[\Delta, \delta_1, \delta_2]^k, [\Delta, \delta_2]^{k+1}, \Delta \Rightarrow \gamma_1, \gamma_2, \Gamma, [\gamma_1, \gamma_2, \Gamma]^k} \; \text{weak-l}}{\Delta, \delta_1, \delta_2 \Rightarrow \gamma_1, \gamma_2, \Gamma} \; \text{weak-l}}$$

Applying recursively the transformations above from top to bottom we obtain a *weak**-free proof in SC$^{[]}_{\mathcal{ALC}}$.

We introduced the *weak** rule to avoid *dispensable contexts* during the bottom-up proof search procedure. Using the strategy suggested above, we only apply the weak rules in order to allow further application of promotional rules. The idea is that we don't need to save unnecessary contexts that are variants of already saved contexts.

Example 4 Consider the fully expanded proof tree Π having sequent 4.3 as root.

$$\exists R.A \sqcap \exists R.B \Rightarrow \exists R.(A \sqcap B) \qquad (4.3)$$

$$\cfrac{\cfrac{\cfrac{\cfrac{\cfrac{[A]^2,[\ldots]^3,B \Rightarrow A,[\ldots]^3,[B]^2 \qquad [A]^2,[\ldots]^3,B \Rightarrow B,[\ldots]^3,[B]^2}{[A]^2,[^{\exists R}A,^{\exists R}B]^3,B \Rightarrow A \sqcap B,[^{\exists R}(A \sqcap B)]^3,[B]^2}\ \text{⊓-r}}{[A]^2,[^{\exists R}A,^{\exists R}B]^3,^{\exists R}B \Rightarrow {}^{\exists R}(A \sqcap B),[^{\exists R}(A \sqcap B)]^3,[B]^2}\ \text{prom-∃}}{[A]^2,^{\exists R}A,^{\exists R}B \Rightarrow {}^{\exists R}(A \sqcap B),[B]^2}\ \text{weak*}}{[^{\exists R}A,^{\exists R}B]^1,A \Rightarrow B,[^{\exists R}(A \sqcap B)]^1}\ \text{f-exch}}{} $$

$$\cfrac{\cfrac{\cfrac{\cfrac{\cfrac{[^{\exists R}A,^{\exists R}B]^1,A \Rightarrow B,[^{\exists R}(A \sqcap B)]^1 \qquad \cfrac{[\ldots]^1,A \Rightarrow A,[\ldots]^1}{}}{[\ldots]^1,A \Rightarrow A \sqcap B,[\ldots]^1}\ \text{⊓-r}}{[^{\exists R}A,^{\exists R}B]^1,^{\exists R}A \Rightarrow {}^{\exists R}(A \sqcap B),[^{\exists R}(A \sqcap B)]^1}\ \text{prom-∃}}{^{\exists R}A,^{\exists R}B \Rightarrow {}^{\exists R}(A \sqcap B)}\ \text{weak*}}{\cfrac{^{\exists R}A,^{\exists R}B \Rightarrow \exists R.(A \sqcap B)}{\cfrac{^{\exists R}A,\exists R.B \Rightarrow \exists R.(A \sqcap B)}{\cfrac{\exists R.A,\exists R.B \Rightarrow \exists R.(A \sqcap B)}{\exists R.A \sqcap \exists R.B \Rightarrow \exists R.(A \sqcap B)}\ \text{⊓-l}}\ \text{∃-l}}\ \text{∃-l}}\ \text{∃-r}}{}$$

We can split Π in three fragments named Π_1, Π_2 and Π_3. The fragments are separated by $weak^*$ and frozen-exchanges. Fragments Π_2 and Π_3 correspond to the two different ways to apply the $weak^*$ in the top-sequent of the fragment Π_1.

$$\Pi_1 \equiv \qquad \cfrac{\cfrac{\cfrac{\cfrac{\cfrac{\overset{\Pi_2}{{}^{\exists R}A,^{\exists R}B \Rightarrow {}^{\exists R}(A \sqcap B)}}{^{\exists R}A,^{\exists R}B \Rightarrow \exists R.(A \sqcap B)}\ \text{∃-r}}{^{\exists R}A,\exists R.B \Rightarrow \exists R.(A \sqcap B)}\ \text{∃-l}}{\exists R.A,\exists R.B \Rightarrow \exists R.(A \sqcap B)}\ \text{∃-l}}{\exists R.A \sqcap \exists R.B \Rightarrow \exists R.(A \sqcap B)}\ \text{⊓-l}$$

$$\Pi_2 \equiv \qquad \cfrac{\cfrac{\overset{\Pi_3}{[\ldots]^1,A \Rightarrow B,[\ldots]^1} \qquad [\ldots]^1,A \Rightarrow A,[\ldots]^1}{[\ldots]^1,A \Rightarrow A \sqcap B,[\ldots]^1}\ \text{⊓-r}}{[^{\exists R}A,^{\exists R}B]^1,^{\exists R}A \Rightarrow {}^{\exists R}(A \sqcap B),[^{\exists R}(A \sqcap B)]^1}\ \text{prom-∃}$$

$$\Pi_3 \equiv \qquad \cfrac{\cfrac{[A]^2,[\ldots]^3,B \Rightarrow A,[\ldots]^3,[B]^2 \qquad [A]^2,[\ldots]^3,B \Rightarrow B,[\ldots]^3,[B]^2}{[A]^2,[\ldots]^3,B \Rightarrow A \sqcap B,[\ldots]^3,[B]^2}\ \text{⊓-r}}{[A]^2,[^{\exists R}A,^{\exists R}B]^3,^{\exists R}B \Rightarrow {}^{\exists R}(A \sqcap B),[^{\exists R}(A \sqcap B)]^3,[B]^2}\ \text{prom-∃}$$

Regarding the contexts created during the proof, contexts 1 and 3 were not turned active yet, they are called auxiliary contexts, they were created during the bottom-up proof construction to save a proof state to further activation and transformation with the system rules, if necessary. Context 1 was used but context 3 was not. Context 2 is the top-sequent of fragment Π_2, saved after been reduced. The idea is that from the fragments Π_2 and Π_3 we can construct a counter-model for the root sequent of Π.

Lemma 7 *If P is a fully expanded proof-tree in $SC^{[]}{}_{\mathcal{ALC}}$ with sequent S as root (conclusion) and if P is in the normal form, from any top-sequent not initial (non-axiom), one can construct a counter-model for S.*

Proof To prove Lemma 7 we must first identify all possible top-sequents in SC$^{[]}{}_{\mathcal{ALC}}$. If weak rules are not allowed during the derivation, all top-sequents in SC$^{[]}{}_{\mathcal{ALC}}$ would have the general form of (4.4).

$$\underbrace{A_1, \ldots, A_n,}_{\Delta_1} \underbrace{{}^{\forall R_1, L_1} B_1, \ldots, {}^{\forall R_m, L_m} B_m}_{\Delta_2} \Rightarrow \underbrace{C_1, \ldots, C_l,}_{\Delta_3} \underbrace{{}^{\exists R_1, L_1} D_1, \ldots, {}^{\exists R_p, L_p} D_p}_{\Delta_4}$$

(4.4)

where we group the concepts into four sets Δ_1, Δ_2, Δ_3 and Δ_4. $A_{1,n}$ and $C_{1,l}$ are sets of atomic concepts. In Δ_2, $B_{1,m}$ are atomic concepts or disjunctions of concepts (not necessarily atomic). In Δ_4, $D_{1,p}$ are atomic concepts or conjunctions of concepts (not necessarily atomic).

To see that no other rule of SC$^{[]}{}_{\mathcal{ALC}}$, rather than weak, could be applied in a sequent like (4.4), one has just to observe that: (1) the ⊓-r and ⊔-l rules provisos are blocking the decomposition of the conjunctions and disjunctions; and (2) the prom-∀ (prom-∃) rule cannot be applied due to the lack of a universal (existential) quantified concept on the right (left).

Nevertheless, with the presence of *weak** rule and considering the strategy for construct normal derivations, *weak** can always be applied to top-sequents like 4.4 reducing them to the simpler cases below. For each one, we will see that it is possible to construct a counter-model.

Case $\Delta_1 \Rightarrow \Delta_3$

That is, a sequent $A_1, \ldots, A_n \Rightarrow C_1, \ldots, C_l$ without labeled concept, it is easy to construct a counter-model $\mathcal{I}$ such that there is an element $a \in (A_1 \sqcap \ldots \sqcap A_n)^{\mathcal{I}}$ and $a \notin (C_1 \sqcup \ldots \sqcup C_l)^{\mathcal{I}}$.

Case $\Delta_2 \Rightarrow$

We can construct a counter-model $\mathcal{I}$ such that there is an element $a \in ({}^{\forall R_1, L_1} B_1 \sqcap \ldots \sqcap {}^{\forall R_m, L_m} B_m)^{\mathcal{I}}$. The right side of a sequent is interpreted as a disjunction, so that, if empty, its semantics for any interpretation function is the empty set. If we consider the simplified case where all roles (labels) are equal, that is ${}^{\forall R, L_1} B_1, \ldots, {}^{\forall R, L_m} B_m \Rightarrow$, we only need to provide a new element a without fillers in R, that is, $\exists x(a, x) \notin R^{\mathcal{I}}$. For the general case, where the most external roles on each concept can be different, the element a cannot have fillers in any of the roles. That is, $\forall R$ occuring in front of the list of labels in Δ_2, $\exists x(a, x) \notin R^{\mathcal{I}}$. We must mention that even if one of the concepts in Δ_2 is ⊤ or ⊥, we can always construct $\mathcal{I}$.

Case $\Rightarrow \Delta_4$

We can construct a counter-model such that $\mathcal{I} \not\models \Rightarrow \Delta_4$. From the natural interpretation of a sequent, we know that an interpretation will not satisfy this case when there is at least one element $a \notin ({}^{\exists R_1, L_1} D_1 \sqcup \ldots \sqcup {}^{\exists R_p, L_p} D_p)^{\mathcal{I}}$. Since the left side of a sequent is interpreted as a conjunction, if empty, its semantics for any interpretation function is the universe set of the interpretation. Once more, let us first consider the case where all existential roles are equal, $^{\exists R,L_1}D_1 \sqcup \ldots \sqcup {}^{\exists R,L_p}D_p$. We only need to provide an element a without fillers in R. If we have different roles in the sequent, a can not have fillers in any of them.

Case $\Delta_2 \Rightarrow \Delta_4$

This case can be reduced for the two cases above. We can always provide an element $a \in \Delta_2^{\mathcal{I}}$ (by second case) and a $\notin \Delta_4^{\mathcal{I}}$ (by third case). In both cases, a will be a fresh element without fillers in any R, for all R most external labels of Δ_2 and Δ_4.   □

Lemma 8 *If P is a weak*-free proof fragment with at least one top-sequent not initial and having S as the bottom sequent, that is, a fragment where the weak rule was not applied. If $\mathcal{I}$ is a counter-model for one of its top-sequents, There is $\mathcal{I}'$ that is a counter-model for S.*

Proof We prove Lemma 8 by case analysis considering each possible rule application and showing how to extend an interpretation that is counter-model of the premiss to be a counter-model of the conclusion.

Cases $\forall$-{l,r} and $\exists$-{l,r}

In these rules the premiss and conclusion have the same semantics, that is, a counter-model for its premiss is also a counter-model for its conclusions.

Cases $\sqcup$-{l,r} and $\sqcap$-{l,r}

Let us first consider the rule $\sqcup$-l. Let $\mathcal{I}$ be an interpretation counter-model for at least one of the premiss. That is, $(\Delta \sqcap {}^{\exists L}\alpha)^{\mathcal{I}} \not\subset \Gamma^{\mathcal{I}}$ or $(\Delta \sqcap {}^{\exists L}\beta)^{\mathcal{I}} \not\subset \Gamma^{\mathcal{I}}$. If any of these cases holds, we have $(\Delta \sqcap {}^{\exists L}\alpha)^{\mathcal{I}} \cup (\Delta \sqcap {}^{\exists L}\beta)^{\mathcal{I}} \not\subset \Gamma^{\mathcal{I}}$ and by the distributivity of the intersection over the union $(\Delta \sqcap ({}^{\exists L}\alpha \sqcup {}^{\exists L}\beta))^{\mathcal{I}} \not\subset \Gamma^{\mathcal{I}}$, which is semantically equivalent to conclusion of the rule: $(\Delta \sqcap ({}^{\exists L}\alpha \sqcup \beta))^{\mathcal{I}} \not\subset \Gamma^{\mathcal{I}}$. Case $\sqcap$-r would be proved in the same way by showing that if $A \not\subset B \cup D$ or $A \not\subset C \cup D$ then $A \not\subset (B \cap C) \cup D$. Rules $\sqcap$-l and $\sqcup$-r are even simpler given the natural interpretation of the sequents. Basically, we are using the results of Sect. 3.2 which shows that these rules are double-sound.

Case $\neg$-l and $\neg$-r

First rule $\neg$-r where δ a labeled concept and $\neg\delta$ its negation. Let us consider a interpretation $\mathcal{I}$ such that $\mathcal{I} \not\models \Delta, \delta \Rightarrow \Gamma$. So we have an element $a \in (\Delta \sqcap \delta)^{\mathcal{I}}$ and $a \notin \Gamma^{\mathcal{I}}$. Thus, $a \in \delta^{\mathcal{I}}$ and so, $a \notin (\neg\delta)^{\mathcal{I}}$. Consequently, $a \notin (\neg\delta \sqcup \Gamma)^{\mathcal{I}}$ as desired. The case of rule $\neg$-l is similar.

Case prom-∃

Assume that we have $\mathcal{I} \not\models \delta \Rightarrow \Gamma$. So we have an element $b \in \delta^{\mathcal{I}}$ and $b \notin \Gamma^{\mathcal{I}}$. We now construct $\mathcal{I}'$ extending $\mathcal{I}$ with one more new element a in the domain and the tuple $(a, b) \in R^{\mathcal{I}}$. In this way, we obtain the necessary condition to $\mathcal{I}' \not\models {}^{+\exists R}\delta \Rightarrow {}^{+\exists R}\Gamma$ which is $a \in {}^{+\exists R}\delta^{\mathcal{I}}$ and $a \notin {}^{+\exists R}\Gamma^{\mathcal{I}}$ since a is a fresh element.

Case prom-∀

Assume that we have $\mathcal{I} \not\models \Delta \Rightarrow \gamma$. Once more, we have an element $b \in \Delta^{\mathcal{I}}$ and $b \notin \gamma^{\mathcal{I}}$. We construct $\mathcal{I}'$ as in the case above, introducing one new element a in the domain and the tuple $(a, b) \in R^{\mathcal{I}}$. Since a is a fresh element with just one filler in R, we guarantee by construction that $a \in {}^{+\forall R}\Delta^{\mathcal{I}}$ and $a \notin {}^{+\forall R}\gamma^{\mathcal{I}}$ and so, $\mathcal{I}' \not\models {}^{+\forall R}\Delta \Rightarrow {}^{+\forall R}\gamma$. Alternatively, we can also introduce in $\mathcal{I}'$ the element a without any filler in R to guarantee that $\mathcal{I}'$ will also be a counter-model for the conclusion.   □

Lemmas 7 and 8 guarantee that from the top-sequents we can construct counter-models and extend them in fragments $weak^*$-free. The following lemma states that we can merge counter-models of proof fragments with top-sequents that are not axioms.

Lemma 9 *Given a weak* application with a conclusion S, reading top-down, this application has two proof fragments with roots S_1 and S_2, their premise and the context that was frozen. If there are interpretations $\mathcal{I}_1$ and $\mathcal{I}_2$ such that $\mathcal{I}_1 \not\models S_1$ and $\mathcal{I}_2 \not\models S_2$ then there is $\mathcal{I}$ such that $\mathcal{I} \not\models S$.*

Proof Without lost of generality, we can consider (4.5) a general format for sequents conclusion of $weak^*$ application. Remember that if we use the strategy define previous, $weak^*$ will only be applied in order to permit promotional rules applications. The case with two existential quantified concepts on the left and two universal quantified concepts on the right will be sufficient to tread all possible combinations. The result of this proof can be easily generalized.

$$\Delta, {}^{\forall R, L_1}\alpha_1, {}^{\exists R, L_2}\alpha_2, {}^{\exists R, L_3}\alpha_3 \Rightarrow \Gamma, {}^{\forall R, L_4}\alpha_4, {}^{\forall R, L_5}\alpha_5, {}^{\exists R, L_6}\alpha_6 \qquad (4.5)$$

To prove Lemma 9, we have to consider each possible pair of proof fragments that a $weak^*$ rule can combine in a top-down construction. In addition, we assume as hypothesis that for both fragments we already constructed a counter-model for its roots—from Lemmas 7 and 8.

1. $S \equiv \Delta, {}^{\exists R, L_2}\alpha_2 \Rightarrow \Gamma, {}^{\exists R, L_6}\alpha_6$. From the hypothesis, we have $\mathcal{I}_1 \not\models \Delta \Rightarrow \Gamma$ and $\mathcal{I}_2 \not\models {}^{\exists R, L_2}\alpha_2 \Rightarrow {}^{\exists R, L_6}\alpha_6$, that is, $\Delta^{\mathcal{I}_1} \not\subseteq \Gamma^{\mathcal{I}_1}$ and ${}^{\exists R, L_2}\alpha_2^{\mathcal{I}_2} \not\subseteq {}^{\exists R, L_6}\alpha_6^{\mathcal{I}_2}$. We create an interpretation $\mathcal{I} = \mathcal{I}_1 \uplus \mathcal{I}_2$, a disjoint union of $\mathcal{I}_1$ and $\mathcal{I}_2$. Now, from $\mathcal{I}_1$ we select an element $a \in \Delta^{\mathcal{I}_1}$ and $a \notin \Gamma^{\mathcal{I}_1}$ that must exist by hypothesis. From $\mathcal{I}_2$ we select an element $b \in \alpha_2^{\mathcal{I}_2}$ and $b \notin \alpha_6^{\mathcal{I}_2}$ that must exist by hypothesis.

Now In $\mathcal{I}$ we add $(a, b) \in R^{\mathcal{I}}$ and we guarantee that $(\Delta \sqcap {}^{\exists R, L_2}\alpha_2)^{\mathcal{I}} \not\subseteq (\Gamma \sqcup {}^{\exists R, L_6}\alpha_6)^{\mathcal{I}}$.

2. $S \equiv \Delta, {}^{\forall R, L_1}\alpha_1 \Rightarrow \Gamma, {}^{\forall R, L_5}\alpha_5$. By hypothesis, we have $\mathcal{I}_1 \not\models \Delta \Rightarrow \Gamma$ and $\mathcal{I}_2 \not\models {}^{\forall R, L_1}\alpha_1 \Rightarrow {}^{\forall R, L_5}\alpha_5$, that is, $\Delta^{\mathcal{I}_1} \not\subseteq \Gamma^{\mathcal{I}_1}$ and ${}^{\forall R, L_1}\alpha_1{}^{\mathcal{I}_2} \not\subseteq {}^{\forall R, L_5}\alpha_5{}^{\mathcal{I}_2}$. We create the interpretation $\mathcal{I}$ as in the previous case, $\mathcal{I} = \mathcal{I}_1 \uplus \mathcal{I}_2$. From $\mathcal{I}_1$ we select an element $a \in \Delta^{\mathcal{I}_1}$ and $a \notin \Gamma^{\mathcal{I}_1}$. From $\mathcal{I}_2$ we select an element $b \in \alpha_1^{\mathcal{I}_2}$ and $b \notin \alpha_5^{\mathcal{I}_2}$. In $\mathcal{I}$ we add $(a, b) \in R^{\mathcal{I}}$ and we guarantee that $(\Delta \sqcap {}^{\forall R, L_1}\alpha_1)^{\mathcal{I}} \not\subseteq (\Gamma \sqcup {}^{\forall R, L_5}\alpha_5)^{\mathcal{I}}$.

3. $S \equiv {}^{\exists R, L_2}\alpha_2, {}^{\exists R, L_3}\alpha_3 \Rightarrow {}^{\exists R, L_6}\alpha_6$. By hypothesis, we have $\mathcal{I}_1 \not\models {}^{\exists R, L_2}\alpha_2 \Rightarrow {}^{\exists R, L_6}\alpha_6$ and $\mathcal{I}_2 \not\models {}^{\exists R, L_3}\alpha_3 \Rightarrow {}^{\exists R, L_6}\alpha_6$. We create the interpretation $\mathcal{I}$ as in the previous case, $\mathcal{I} = \mathcal{I}_1 \uplus \mathcal{I}_2$. From $\mathcal{I}_1$ we have $a \in ({}^{\exists R, L_2}\alpha_2)^{\mathcal{I}_1}$, and thus, an $(a, b) \in R^{\mathcal{I}_1}$ with $b \in \alpha_2^{\mathcal{I}_1}$. From $\mathcal{I}_2$ we have $b \in ({}^{\exists R, L_3}\alpha_3)^{\mathcal{I}_2}$, and thus, an $(b, c) \in R^{\mathcal{I}_2}$ with $c \in \alpha_3^{\mathcal{I}_2}$. We create now a fresh element d and add in $R^{\mathcal{I}}$ the set $\{(d, b), (d, c)\}$. We have guarantee that $d \in ({}^{\exists R, L_2}\alpha_2 \sqcap {}^{\exists R, L_3}\alpha_3)^{\mathcal{I}}$ and $d \notin ({}^{\exists R, L_6}\alpha_6)^{\mathcal{I}}$. Note that $b \notin ({}^{\exists R, L_6}\alpha_6)^{\mathcal{I}}$ (resp. c) by hypothesis.

4. If we consider $\forall R.\alpha \equiv \neg \exists R.\neg \alpha$, cases $S \equiv {}^{\exists R, L_2}\alpha_2, {}^{\forall R, L_1}\alpha_1 \Rightarrow {}^{\exists R, L_6}\alpha_6$, ${}^{\forall R, L_4}\alpha_4$ and $S \equiv {}^{\forall R, L_1}\alpha_1 \Rightarrow {}^{\forall R, L_4}\alpha_4, {}^{\forall R, L_5}\alpha_5$ has been already considered. $\square$

Reference

1. Baader, F.: The Description Logic Handbook: Theory, Implementation, and Applications. Cambridge University Press, Cambridge (2003)

Chapter 5
A Natural Deduction for $\mathcal{ALC}$

Abstract In this chapter we present a Natural Deduction (**ND**) system for $\mathcal{ALC}$, named ND$_{\mathcal{ALC}}$. We briefly discuss the motivation and the basic considerations behind the design of ND$_{\mathcal{ALC}}$. We also prove the completeness, soundness and the normalization theorem for ND$_{\mathcal{ALC}}$.

Keywords Natural deduction · $\mathcal{ALC}$ · Normalization · Completeness · Soundness · Proof theory

5.1 Introduction

It is quite well-known the fact that Natural Deduction (**ND**) proofs in intuitionistic logic (**IL**) have computational content. This content can be explicitly read from the typed λ-calculus term associated to each proof. Moreover, to each normalization step that can be applied in the proof, there is a corresponding β-reduction in its associated typed λ-term. This is known as the Curry-Howard isomorphism (**CH-ISO**) between **ND** and the typed λ-calculus [3]. For classical logic this isomorphism does not hold any more. However, there are some attempts to justify weak or modified forms of this isomorphism for classical logic (see [1, 2] for example).

It seems to exist some connections between the computational content of a proof and its ability to provide good structures to explanation extraction from proofs. In fact, an algorithm is one of the most precise arguments to explain how to obtain a result out of some inputs. Given that, translating algorithms according the propositions-as-types **CH-ISO** we should obtain a quite good argument establishing the conclusion from the premises. Despite the fact that for classical logic the **CH-ISO** is not well-established at all, we still argue in favor of ND proofs instead of Sequent Calculus (**SC**) in order to provide good explanations. One of the main points in favor of **ND** is the fact that it is single-conclusion and provides, in this way, a direct chain of inferences linking the propositions in the proof. It is worth noting that there is more

$$\frac{L^\forall (\alpha \sqcap \beta)}{L^\forall \alpha} \ \sqcap\text{-e} \qquad \frac{L^\forall (\alpha \sqcap \beta)}{L^\forall \beta} \ \sqcap\text{-e} \qquad \frac{L^\forall \alpha \quad L^\forall \beta}{L^\forall (\alpha \sqcap \beta)} \ \sqcap\text{-i}$$

$$\frac{L^\exists (\alpha \sqcup \beta) \quad \overset{[L^\exists \alpha]}{\overset{\vdots}{\gamma}} \quad \overset{[L^\exists \beta]}{\overset{\vdots}{\gamma}}}{\gamma} \ \sqcup\text{-e} \qquad \frac{L^\exists \alpha}{L^\exists (\alpha \sqcup \beta)} \ \sqcup\text{-i} \qquad \frac{L^\exists \beta}{L^\exists (\alpha \sqcup \beta)} \ \sqcup\text{-i}$$

$$\frac{L\alpha \quad \neg L \neg \alpha}{\bot} \ \neg\text{-e} \qquad \frac{\overset{[L\alpha]}{\overset{\vdots}{\bot}}}{\neg L \neg \alpha} \ \neg\text{-i} \qquad \frac{\overset{[\neg L \neg \alpha]}{\overset{\vdots}{\bot}}}{L\alpha} \ \bot_c$$

$$\frac{L \exists R.\alpha}{L, \exists R_\alpha} \ \exists\text{-e} \qquad \frac{L, \exists R_\alpha}{L \exists R.\alpha} \ \exists\text{-i} \qquad \frac{L \forall R.\alpha}{L, \forall R_\alpha} \ \forall\text{-e}$$

$$\frac{L, \forall R_\alpha}{L \forall R.\alpha} \ \forall\text{-i} \qquad \frac{L_1 \alpha \quad L_1 \alpha \sqsubseteq L_2 \beta}{L_2 \beta} \ \sqsubseteq\text{-e} \qquad \frac{\overset{[L_1 \alpha]}{\overset{\vdots}{L_2 \beta}}}{L_1 \alpha \sqsubseteq L_2 \beta} \ \sqsubseteq\text{-i}$$

$$\frac{L\alpha}{\forall R, L_\alpha} \ Gen$$

Fig. 5.1 The natural deduction system for $\mathcal{ALC}$

than one **ND** normal proof related to the same cut-free **SC** proof. It is mainly because of this fact that a (cut-free) **SC** proof is related to more than one **ND** proof. We believe that explanations should be as specific as their proof-theoretical counterparts.

5.2 The ND$_{\mathcal{ALC}}$ System

Figure 5.1 shows the system called ND$_{\mathcal{ALC}}$. Despite the use of labeled formulas, the main non-standard feature of ND$_{\mathcal{ALC}}$ is the fact that it is defined on two kind of "formulas", namely *concept formulas* and *subsumptions of concepts*.

If $\Phi_1, \Phi_2 \vdash \Psi$ is an inference rule involving only concept formulas then it states that whenever the premises are taken as non-empty collections of individuals the conclusion is taken as non-empty too. Particularly, providing any DL-interpretation for the premise concepts, if a is an individual belonging to both interpreted concepts then it also belongs to the interpreted conclusion. On the other hand, a subsumption $\Phi \sqsubseteq \Psi$ has no concept associate to it. It states, instead, a truth-value statement, depending on whether the interpretation of Φ is included in the corresponding interpretation of Ψ. In terms of a logical system, DL has no concept internalizing $\sqsubseteq$.

As we will see on the next section, this imposes quite particular features on the form of the normal proofs in ND$_{\mathcal{ALC}}$.

In the rule $\sqsubseteq$-i, $^{L_1}\alpha \sqsubseteq {}^{L_2}\beta$ only depends on the assumption $^{L_1}\alpha$ and no other hypothesis. The proviso to the application of rule *Gen* application is that the premise $^L\alpha$ does not depend on any hypothesis. In $\bot_c$-rule, $^L\alpha$ has to be different from $\bot$. In some rules the list of labels L has a superscript, $L^\forall$ or $L^\exists$. This notation means that the list of labels L should contain only $\forall R$ (resp. $\exists R$) labels. When L has no superscript, any kind of label is allowed.

The semantics of ND$_{\mathcal{ALC}}$ follows the $\mathcal{ALC}$ semantics presented in Sect. 2.1, that is, it is given by an *interpretation*. However, since ND$_{\mathcal{ALC}}$ deals with two different kind of formulas, we must define how an interpretation satisfies both kinds.

Definition 13 Let $\Omega = (\mathcal{C}, \mathcal{S})$ be a tuple composed by a set of labeled concepts $\mathcal{C} = \{\alpha_1, \ldots, \alpha_n\}$ and a set of subsumption $\mathcal{S} = \{\gamma_1^1 \sqsubseteq \gamma_2^1, \ldots, \gamma_1^k \sqsubseteq \gamma_2^k\}$. We say that an interpretation $\mathcal{I} = (\Delta^{\mathcal{I}}, \cdot^{\mathcal{I}})$ satisfies Ω and write $\mathcal{I} \models \Omega$ whenever:

1. $\mathcal{I} \models \mathcal{C}$, which means $\bigcap_{\alpha \in \mathcal{C}} \sigma(\alpha)^{\mathcal{I}} \neq \emptyset$; and
2. $\mathcal{I} \models \mathcal{S}$, which means that for all $\gamma_1^i \sqsubseteq \gamma_2^i \in \mathcal{S}$, we have $\sigma(\gamma_1^i)^{\mathcal{I}} \subseteq \sigma(\gamma_2^i)^{\mathcal{I}}$.

We adopted the standard notation $\Omega \vdash F$ if there is a deduction Π with conclusion F (concept or subsumption) from Ω as set of hypothesis.

5.3 ND$_{\mathcal{ALC}}$ Soundness

Lemma 10 *Let Π be a deduction in* ND$_{\mathcal{ALC}}$ *of F with all hypothesis in $\Omega = (\mathcal{C}, \mathcal{S})$, then if F is a concept:*

$$\mathcal{S} \models \left(\prod_{A \in \mathcal{C}} A\right) \sqsubseteq F$$

and if F is a subsumption $A_1 \sqsubseteq A_2$:

$$\mathcal{S} \models \left(\prod_{A \in \mathcal{C}} A\right) \sqcap A_1 \sqsubseteq A_2$$

For sake of clear presentation in the following proof we adopt some special notations. We will write $\forall L.\alpha$ to abbreviate $\forall R_1. \ldots .\forall R_n.\alpha$ when $L = \forall R_1. \ldots .\forall R_n$. The labeled concept $^L\alpha$ will be taken as equivalent to its $\mathcal{ALC}$ correspondent concept $\sigma(^L\alpha)$.[1] Letters γ and δ stand for labeled concepts while α and β stand for $\mathcal{ALC}$ concepts. We take $\mathcal{C}$ as $\prod_{A \in \mathcal{C}} A$. We will aso use many times the axioms presented in Sect. 2.6.

Proof The proof of Lemma 10 is done by induction on the height of the proof tree Π, represented by $|\Pi|$.

---

[1] In Sect. 3.1 the reader can find the definition of σ function and labeled formulas.

Base case

If $\mid \Pi \mid = 1$ then $\Omega \vdash {}^L\alpha$ is such that ${}^L\alpha$ is in Ω. In that case, is easy to see that Lemma 10 holds since by basic set theory $(A \cap B) \subseteq A$ for all A and B.

Rule $\sqcap$-e

$$\Pi_1$$

By induction hypothesis, if ${}^L(\alpha \sqcap \beta)$ is a derivation with all hypotheses in $\{\mathcal{C}, \mathcal{S}\}$ then $\mathcal{S} \models \mathcal{C} \sqsubseteq {}^L(\alpha \sqcap \beta)$. From the definition of labeled concepts and Axiom 2.1 we can rewrite to $\mathcal{S} \models \mathcal{C} \sqsubseteq {}^L\alpha \sqcap {}^L\beta$ which from basic set theory guarantee $\mathcal{S} \models \mathcal{C} \sqsubseteq {}^L\alpha$.

Rule $\sqcap$-i

$$\Pi_1 \qquad \Pi_2$$

Let us consider the two derivations ${}^L\alpha$ and ${}^L\beta$ with all hypothesis in $\{\mathcal{C}_1, \mathcal{S}_1\}$ and $\{\mathcal{C}_2, \mathcal{S}_2\}$. By induction hypothesis, (1) $\mathcal{S}_1 \models \mathcal{C}_1 \sqsubseteq {}^L\alpha$ an (2) $\mathcal{S}_2 \models \mathcal{C}_2 \sqsubseteq {}^L\beta$. Now let us consider the deduction

$$\frac{\begin{matrix} \Pi_1 & \Pi_2 \\ {}^L\alpha & {}^L\beta \end{matrix}}{{}^L(\alpha \sqcap \beta)}$$

with all hypothesis in $\{\mathcal{C}_1 \cup \mathcal{C}_2, \mathcal{S}_1 \cup \mathcal{S}_2\}$. It is easy to see that from (1) and (2) $\mathcal{S}_1 \cup \mathcal{S}_2 \models (\mathcal{C}_1 \sqcap \mathcal{C}_2) \sqsubseteq {}^L\alpha$ and $\mathcal{S}_1 \cup \mathcal{S}_2 \models (\mathcal{C}_1 \sqcap \mathcal{C}_2) \sqsubseteq {}^L\beta$. From basic set theory we may write $\mathcal{S}_1 \cup \mathcal{S}_2 \models (\mathcal{C}_1 \sqcap \mathcal{C}_2) \sqsubseteq {}^L\alpha \sqcap {}^L\beta$ and finally from Axiom 2.1 we get the desired result $\mathcal{S}_1 \cup \mathcal{S}_2 \models (\mathcal{C}_1 \sqcap \mathcal{C}_2) \sqsubseteq {}^L(\alpha \sqcap \beta)$.

Rules $\sqcup$-i

$$\Pi_1$$

Again by induction hypothesis, if ${}^L\alpha$ is a derivation with all hypothesis in $\{\mathcal{C}, \mathcal{S}\}$ then $\mathcal{S} \models \mathcal{C} \sqsubseteq {}^L\alpha$. Using basic set theory we can rewrite to $\mathcal{S} \models \mathcal{C} \sqsubseteq {}^L\alpha \sqcup {}^L\beta$ and using Axiom 2.3 to $\mathcal{S} \models \mathcal{C} \sqsubseteq {}^L(\alpha \sqcup \beta)$.

Rule ($\sqcup$-e)

By induction hypothesis, if

$$\begin{matrix} & [{}^L\alpha] & [{}^L\beta] \\ \Pi_1 & \Pi_2 & \Pi_3 \\ {}^L(\alpha \sqcup \beta), & \gamma & \text{and} \quad \gamma \end{matrix}$$

are derivations with hypothesis in $\{\mathcal{C}, \mathcal{S}\}$, $\{{}^L\alpha, \mathcal{S}\}$ and $\{{}^L\beta, \mathcal{S}\}$, respectively. Then, $\mathcal{S} \models \mathcal{C} \sqsubseteq {}^L(\alpha \sqcup \beta)$, $\mathcal{S} \models {}^L\alpha \sqsubseteq \gamma$ and $\mathcal{S} \models {}^L\beta \sqsubseteq \gamma$. From set theory

$\mathcal{S} \models (^L\alpha \sqcup {}^L\beta) \sqsubseteq \gamma$ and from Axiom 2.3, $\mathcal{S} \models {}^L(\alpha \sqcup \beta) \sqsubseteq \gamma$. Now by the transitivity of set inclusion we can get the desired result $\mathcal{S} \models \mathcal{C} \sqsubseteq \gamma$.

Rules ∀-i, ∀-e, ∃-i and ∃-e

They are sound since the premises and conclusions have the same semantics.

Rule ¬-e

By induction hypothesis, if

$$\begin{array}{cc} \Pi_1 & \Pi_2 \\ {}^L\alpha & \text{and} \quad {}^{\neg L}\neg\alpha \end{array}$$

are derivation with hypothesis in $\{\mathcal{C}_1, \mathcal{S}_1\}$ and $\{\mathcal{C}_2, \mathcal{S}_2\}$ we know that $\mathcal{S}_1 \models \mathcal{C}_1 \sqsubseteq {}^L\alpha$ and $\mathcal{S}_2 \models \mathcal{C}_2 \sqsubseteq {}^{\neg L}\neg\alpha$. Now consider the deduction

$$\frac{\begin{array}{cc} \Pi_1 & \Pi_2 \\ {}^L\alpha & {}^{\neg L}\neg\alpha \end{array}}{\bot}$$

with hypothesis in $\{\mathcal{S}_1 \cup \mathcal{S}_2, \mathcal{C}_1 \cup \mathcal{C}_2\}$. By inductive hypothesis we can write $\mathcal{S}_1 \cup \mathcal{S}_2 \models \mathcal{C}_1 \sqsubseteq {}^L\alpha$ and $\mathcal{S}_2 \cup \mathcal{S}_2 \models \mathcal{C}_2 \sqsubseteq {}^{\neg L}\neg\alpha$. Now, from the fact that $\mathcal{ALC}$ semantics states ${}^L\alpha$ and ${}^{\neg L}\neg\alpha$ as two disjoint sets, we have $\mathcal{C}_1 \sqcap \mathcal{C}_2 = \emptyset$ and we can write $\mathcal{S}_1 \cup \mathcal{S}_2 \models (\mathcal{C}_1 \sqcap \mathcal{C}_2) \sqsubseteq \bot$ as desired.

Rule ¬-i

$$\begin{array}{c} {}^L\alpha \\ \Pi_2 \end{array}$$

If $\{\mathcal{C}, \mathcal{S}\}$ holds all the hypothesis of the deduction $\bot$ then by induction hypothesis $\mathcal{S} \models \mathcal{C} \sqcap {}^L\alpha \sqsubseteq \bot$ (taking $\bot$ as its semantics counterpart, namely, the empty set). From basic set theory $\mathcal{S} \models \mathcal{C} \sqsubseteq {}^{\neg L}\neg\alpha$ as desired.

Rule $\bot_c$

The argument is similar from above.

Rule ⊑-e

$$\begin{array}{cc} \Pi_1 & \Pi_2 \end{array}$$

By induction hypothesis, if γ and $\gamma \sqsubseteq \delta$ are deduction with hypothesis in $\{\mathcal{C}_1, \mathcal{S}_1\}$ and $\{\mathcal{C}_2, \mathcal{S}_2\}$, we have (1) $\mathcal{S}_1 \models \mathcal{C}_1 \sqsubseteq \gamma$ and (2) $\mathcal{S}_2 \models \mathcal{C}_2 \sqcap \gamma \sqsubseteq \delta$. Let us now consider the application of rule ⊑-e to construct the derivation

$$\frac{\overset{\Pi_1}{\gamma} \quad \overset{\Pi_2}{\gamma \sqsubseteq \delta}}{\delta}$$

with hypothesis in $\{C_1 \cup C_2, \, S_1 \cup S_2\}$. From (2) and $\mathcal{ALC}$ semantics we can conclude $S_1 \cup S_2 \models C_2 \sqcap \gamma \sqsubseteq \delta$. Finally, from basic set theory $C_1 \sqcap C_2 \sqsubseteq C_2$ we obtain $S_1 \cup S_2 \models C_1 \sqcap C_2 \sqsubseteq \delta$.

Rule $\sqsubseteq$-i

$$\overset{\gamma}{\underset{\Pi_1}{}}$$

By induction hypothesis, if $\overset{}{\delta}$ is a deduction with hypothesis in $\{C, S\}$ then $S \models C \sqsubseteq \delta$ and we conclude $S \models C^- \sqcap \gamma \sqsubseteq \delta$ where C^- is $C - \{\gamma\}$.

Rule *Gen*

Let Π be a proof of $^L\alpha$ following from an empty set of hypothesis, we may write $\vdash {}^L\alpha$. That is, $^L\alpha$ is a DL-tautology or $\sigma(^L\alpha)^{\mathcal{I}} \equiv \Delta^{\mathcal{I}}$. From the necessitation rule from Sect. 2.6, whenever a concept C is a DL-tautology, for any given R, the concept $\forall R.C$ will be also. For that, we can conclude that $^{\forall R, L}\alpha$ for any given R will be also a tautology. Remember that $^{\forall R, L}\alpha \equiv \forall R.\sigma(^L\alpha)$. $\Box$

Let us now state the main theorem of this section.

Theorem 3 ND$_{\mathcal{ALC}}$ *is sound regarding the standard semantics of* $\mathcal{ALC}$.

$$\text{if } \Omega \vdash \gamma \quad \text{then} \quad \Omega \models \gamma$$

where $\Omega = (C, S)$) *is a tuple composed by a set of labeled concepts (C) and subsumptions (S).*

Proof It follows directly from Lemma 10. $\Box$

5.4 ND$_{\mathcal{ALC}}$ Completeness

We use the same strategy from Sect. 3.3 to prove ND$_{\mathcal{ALC}}$ completeness. That is, we show how the axiomatic presentation of $\mathcal{ALC}$ can be derived in ND$_{\mathcal{ALC}}$.

Theorem 4 ND$_{\mathcal{ALC}}$ *is complete regarding the standard semantics of* $\mathcal{ALC}$.

Proof The DL rule of generalization

$$\frac{\vdash \alpha}{\vdash \forall R.\alpha}$$

is a derived rule of ND$_{\mathcal{ALC}}$, for supposing $\vdash \alpha$ implies the existence of a proof (without hypothesis) Π of α. We prove $\forall R.\alpha$, without any new hypothesis by means of the following schema:

$$
\begin{array}{c}
\Pi \\
\vdots \\
\dfrac{\alpha}{\dfrac{R_\alpha}{\forall R.\alpha}\ \text{$\forall$-i}}\ Gen
\end{array}
$$

The following proofs justifies in ND$_{\mathcal{ALC}}$ the $\mathcal{ALC}$ axiom $\forall R.(A \sqcap B) \equiv (\forall R.A \sqcap \forall R.B)$, where $\alpha \equiv \beta$ is an abbreviation for $\alpha \sqsubseteq \beta$ and $\beta \sqsubseteq \alpha$, having obvious $\equiv$ elimination and introduction rules, based on $\sqsubseteq$ elimination and introduction rules.

$$
\dfrac{\dfrac{\dfrac{\dfrac{[\forall R.(A \sqcap B)]}{\forall R(A \sqcap B)}\ \text{$\forall$-e}}{\dfrac{\forall R A}{\forall R.A}\ \text{$\forall$-i}}\ \text{$\sqcap$-e} \qquad \dfrac{\dfrac{\dfrac{[\forall R.(A \sqcap B)]}{\forall R(A \sqcap B)}\ \text{$\forall$-e}}{\dfrac{\forall R B}{\forall R.B}\ \text{$\forall$-i}}\ \text{$\sqcap$-e}}{\forall R.A \sqcap \forall R.B}\ \text{$\sqcap$-i}}{\forall R.(A \sqcap B) \sqsubseteq \forall R.A \sqcap \forall R.B}\ \text{$\sqsubseteq$-i}
$$

$$
\dfrac{\dfrac{\dfrac{\dfrac{[\forall R.A \sqcap \forall R.B]}{\forall R.A}\ \text{$\sqcap$-e}}{\forall R A}\ \text{$\forall$-e} \qquad \dfrac{\dfrac{[\forall R.A \sqcap \forall R.B]}{\forall R.B}\ \text{$\sqcap$-e}}{\forall R B}\ \text{$\forall$-e}}{\dfrac{\forall R(A \sqcap B)}{\forall R.(A \sqcap B)}\ \text{$\forall$-i}}\ \text{$\sqcap$-i}}{\forall R.A \sqcap \forall R.B \sqsubseteq \forall R.(A \sqcap B)}\ \text{$\sqsubseteq$-i}
$$

ND$_{\mathcal{ALC}}$ is a conservative extension of the classical propositional calculus. To see that, let Δ be a set of formulas of the form $\{\gamma_1, \ldots, \gamma_k, \alpha_1 \rightarrow \beta_1, \ldots, \alpha_n \rightarrow \beta_n\}$, where each γ_i, α_i and β_i are propositional formulas and α_i and β_i do not have any occurrence of $\rightarrow$. One can easily verify that any propositional classical consequence $\Delta \models \alpha$ is justified by a proof in classical **ND**. Now trasform this proof into a proof in ND$_{\mathcal{ALC}}$ by replacing each $\rightarrow$ by $\sqsubseteq$.

Since ND$_{\mathcal{ALC}}$ is a conservative extension of the classical propositional **ND** system that has the generalization as a derived rule, and, proves axiom $\forall R.(A \sqcap B) \equiv (\forall R.A \sqcap \forall R.B)$, we have the completeness for ND$_{\mathcal{ALC}}$ by a relative completeness to the axiomatic presentation of $\mathcal{ALC}$. $\qquad\qquad\square$

5.5 Normalization Theorem for ND$_{\mathcal{ALC}}$

In this section we prove the normalization theorem for ND$_{\mathcal{ALC}}$. It is worth nothing that the usual reductions for obtaining a normal proof in classical propositional logic also applies to ND$_{\mathcal{ALC}}$. Thus, the first thing to observe is that we follow Prawitz's [4] approach incremented by Seldin's [5] permutation rules for the classical absurdity $\perp_c$. That is, using a set of permutative rules, we move any application of $\perp_c$-rule downwards the conclusion. After this transformation we end up with a proof having in each *branch* at most one $\perp_c$-rule application as the last rule of it.

In order to move the absurdity rule downwards the conclusion and also to have a more succinct proof we restrict the language to the fragment $\{\neg, \forall, \sqcap, \sqsubseteq\}$. This will ·not limit our results since any $\mathcal{ALC}$ formula can be rewritten in an equivalent one in this restricted fragment. We shall consider the system $\text{ND}^-_{\mathcal{ALC}}$ obtained from $\text{ND}_{\mathcal{ALC}}$ by removing from $\text{ND}_{\mathcal{ALC}}$ $\sqcup$-rules and $\exists$-rules. The Proposition 1 states that the system $\text{ND}^-_{\mathcal{ALC}}$ is essentially just a syntactic variation of $\text{ND}_{\mathcal{ALC}}$ system.

Proposition 1 *The* $\text{ND}_{\mathcal{ALC}}$ $\sqcup$*-rules and* $\exists$*-rules are derived in* $\text{ND}^-_{\mathcal{ALC}}$.

Proof Considering the concept description $^L\alpha \sqcup \beta$ being defined by $^L\neg(\neg\alpha \sqcap \neg\beta)$ and the concept description $^L\exists R.\alpha$ by $^L\neg\forall R.\neg\alpha$.

The rules ($\sqcup$-i) can be derived as follows:

$$
\dfrac{{}^L\alpha \quad \dfrac{[^{\neg L}(\neg\alpha \sqcap \neg\beta)]^1}{\neg{}^L\neg\alpha}\ \text{\footnotesize$\sqcap$-e}}{\dfrac{\perp}{{}^L\neg(\neg\alpha \sqcap \neg\beta)}\ \text{\footnotesize$\neg$-i}}\ \text{\footnotesize$\neg$-e}
\qquad\qquad
\dfrac{{}^L\beta \quad \dfrac{[^{\neg L}(\neg\alpha \sqcap \neg\beta)]^1}{\neg{}^L\neg\beta}\ \text{\footnotesize$\sqcap$-e}}{\dfrac{\perp}{{}^L\neg(\neg\alpha \sqcap \neg\beta)}\ \text{\footnotesize$\neg$-i}}\ \text{\footnotesize$\neg$-e}
$$

where L contains only existencial quantified labels. $\neg L$ as described in Sect. 3.1, is the negation of L, that is, universal quantifiers are changed to existential quantifiers and vice-versa. We note that rule $\sqcup$-i proviso requires that L only contains existential quantified labels, what makes the rule $\sqcap$-e proviso satisfied since $\neg L$ will only contain universal quantified labels. The rule $\sqcup$-e can also be derived:

$$
\dfrac{\dfrac{\dfrac{\begin{array}{cc}[^L\alpha] & \quad [^L\beta]\\ \vdots & \vdots \\ \dot\gamma\quad[\neg\gamma] & \dot\gamma\quad[\neg\gamma]\\ \dfrac{\perp}{\neg{}^L\neg\alpha} & \dfrac{\perp}{\neg{}^L\neg\beta}\end{array}}{\neg{}^L(\neg\alpha\sqcap\neg\beta)} \qquad {}^L\neg(\neg\alpha\sqcap\neg\beta)}{\perp}}{\gamma}
$$

For rules ($\exists$-i) and ($\exists$-e), it is worth noting that $\text{ND}^-_{\mathcal{ALC}}$ does not restrict the occurrence of existential labels, only the existential constructor of $\mathcal{ALC}$. In other words, we have just reused the $\mathcal{ALC}$ constructors $\forall$ and $\exists$ to "type" the labels and keep track of the original role quantification when it is promoted to label. Nevertheless, although the confusion could be avoided if we adopted $\neg\forall R$ instead of $\exists R$ in the labels of $\text{ND}^-_{\mathcal{ALC}}$ concepts, for clear presentation we choose to allow $\exists R$ on $\text{ND}^-_{\mathcal{ALC}}$ concept's labels. $\square$

$$
\begin{array}{cc}
\cfrac{
{}^{L}{,}\exists R_\alpha \quad
\cfrac{[{}^{\neg L}\forall R.\neg\alpha]}{(\neg L),\forall R_{\neg\alpha}}
}{
\cfrac{\bot}{{}^{L}{}_{\neg}\forall R.\neg\alpha}
}
& \quad
\cfrac{
\cfrac{[{}^{(\neg L),\forall R}{}_{\neg\alpha}]}{{}^{\neg L}\forall R.\neg\alpha} \quad {}^{L}{}_{\neg}\forall R.\neg\alpha
}{
\cfrac{\bot}{{}^{L}{,}\exists R_\alpha}
}
\end{array}
$$

In the sequel, we adopt Prawitz's [4] terminologies such as: formula-tree, deductions or derivations, rule application, minor and major premises, *threads, branches* and so on. Nevertheless some terminologies have different definition in our system, in that case, we will present that definition.

A *branch* in a ND$_{\mathcal{ALC}}$ or ND$^-{}_{\mathcal{ALC}}$ deduction is an initial part $\alpha_1, \alpha_2, \ldots, \alpha_n$ of a thread such that α_n is either (i) the first formula occurrence in the thread that is a minor premise of an application of $\sqsubseteq$-e or (ii) the last formula occurrence of a thread (the end-formula of the deduction) if there is no such premise in the thread.

Given a deduction Π on ND$_{\mathcal{ALC}}$ or ND$^-{}_{\mathcal{ALC}}$, we define the *height* of a formula occurrence α in Π inductively:

- if α is the end-formula of Π (conclusion), then $h(\alpha) = 0$;
- if α is a premise of a rule application, say λ, in Π and is not the end-formula of Π, then $h(\alpha) = h(\beta) + 1$ where β is the conclusion of λ.

In a similar matter we can define the height of a *rule application* in a deduction.

A *maximal formula* is a formula occurrence that is consequence of an introduction rule and the major premise of an elimination rule. A maximal $\sqsubseteq$-formula in a proof Π is a maximal formula that is a subsumption.

Lemma 11 *Let Π be a proof of α (concept or subsumption of concepts) from Δ in* ND$^-{}_{\mathcal{ALC}}$. *Then there is a proof Π' without maximal $\sqsubseteq$-formulas.*

Proof We prove Lemma 11 by induction over the number of maximal $\sqsubseteq$-formulas occurrences. We apply a sequence of reductions choosing always a highest maximal $\sqsubseteq$-formula occurence in the proof tree. In the reduction shown below we note that α cannot be a subsumpption, so that, the reduction application will never introduce new maximal $\sqsubseteq$-formulas. In other words, we cannot have nested subsumptions, subsumptions are not concepts. $\square$

$$
\cfrac{\Pi_1 \quad \cfrac{[\alpha] \\ \Pi_2 \\ \beta}{\alpha \sqsubseteq \beta}}{\beta}
\quad \triangleright \quad
\begin{array}{c} \Pi_1 \\ [\alpha] \\ \Pi_2 \\ \beta \end{array}
$$

Lemma 12 (Moving $\bot_c$ downwards on branches) *If $\Omega \vdash_{\text{ND}^-{}_{\mathcal{ALC}}} \alpha$, then there is a deduction Π in* ND$^-{}_{\mathcal{ALC}}$ *of α from Ω where each branch in Π has at most one application of $\bot_c$-rule and, whenever it has one, it is one of the following cases: (i) the last rule applied in this branch; (ii) its conclusion is the premise of a $\sqsubseteq$-i application, being this $\sqsubseteq$-i the last rule applied in the branch.*

Proof Let Π be a deduction in $\text{ND}^-\mathcal{ALC}$ of α (subsumption of concepts or concept) from a set of hypothesis Δ. Let λ be an application of a $\perp_c$-rule in Π with $h(\lambda) = d$ such that there is no other application of $\perp_c$-rule above λ. Let us consider each possible rule application immediately below λ. For each case, we show how one can exchange the rules decreasing the height of λ.

Rule $\forall$-e

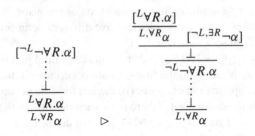

$$
\cfrac{[^{\neg L}\neg\forall R.\alpha]}{\underset{\vdots}{}\;\;\cfrac{\perp}{\cfrac{{}^L\forall R.\alpha}{L,\forall R_\alpha}}}
\qquad \triangleright \qquad
\cfrac{\cfrac{\cfrac{[^L\forall R.\alpha]}{L,\forall R_\alpha}\quad[^{\neg L,\exists R}\neg\alpha]}{\perp}}{\cfrac{\neg^L\neg\forall R.\alpha}{\underset{\vdots}{}\;\cfrac{\perp}{L,\forall R_\alpha}}}
$$

Rule $\forall$-i

$$
\cfrac{[^{\neg L,\exists R}\neg\alpha]}{\underset{\vdots}{}\;\;\cfrac{\perp}{\cfrac{L,\forall R_\alpha}{{}^L\forall R.\alpha}}}
\qquad \triangleright \qquad
\cfrac{\cfrac{\cfrac{[L,\forall R_\alpha]}{{}^L\forall R.\alpha}\quad[^{\neg L}\neg\forall R.\alpha]}{\perp}}{\cfrac{\neg L,\exists R_{\neg\alpha}}{\underset{\vdots}{}\;\cfrac{\perp}{{}^L\forall R.\alpha}}}
$$

Rule $\sqcap$-i

$$
\cfrac{\exists L_{\neg\alpha}}{\underset{\vdots}{}\;\;\cfrac{\perp}{\cfrac{{}^{\forall L}\alpha\quad \cfrac{\Pi}{{}^{\forall L}\beta}}{{}^{\forall L}(\alpha\sqcap\beta)}}}
\qquad \triangleright \qquad
\cfrac{\cfrac{\cfrac{[^{\forall L}\alpha]^2\quad\cfrac{\Pi}{{}^{\forall L}\beta}}{{}^{\forall L}(\alpha\sqcap\beta)}\quad[^{\exists L}\neg(\alpha\sqcap\beta)]^1}{\perp}\;2}{\cfrac{\exists L_{\neg\alpha}}{\underset{\vdots}{}\;\cfrac{\perp}{{}^{\forall L}(\alpha\sqcap\beta)}\;1}}
$$

Rule ⊓-e

$$
\cfrac{
\begin{array}{c}
{}^{\exists L}\neg(\alpha \sqcap \beta) \\
\vdots \\
\bot \\
\hline
{}^{\forall L}(\alpha \sqcap \beta)
\end{array}
}{
{}^{\forall L}\alpha
}
\qquad \triangleright \qquad
\cfrac{
\begin{array}{c}
[{}^{\exists L}\neg\alpha]^2 \quad \cfrac{[{}^{\forall L}(\alpha \sqcap \beta)]^1}{{}^{\forall L}\alpha} \\
\hline
\cfrac{\bot}{{}^{\exists L}\neg(\alpha \sqcap \beta)} \ 1 \\
\vdots \\
\bot
\end{array}
}{
{}^{\forall L}\alpha
} \ 2
$$

Rule ¬-e

$$
\cfrac{
\begin{array}{c}
[{}^{\neg L}\neg\alpha] \\
\vdots \\
\cfrac{\bot}{{}^{L}\alpha} \quad {}^{\neg L}\neg\alpha \\
\hline
\bot
\end{array}
}{}
\qquad \triangleright \qquad
\cfrac{
\begin{array}{c}
[{}^{L}\alpha] \quad \cfrac{[\Delta] \quad \Pi}{{}^{\neg L}\neg\alpha} \\
\hline
\cfrac{\bot}{{}^{\neg L}\neg\alpha} \\
\vdots \\
\bot
\end{array}
}{}
$$

One must observe that in all reductions above, the conclusion of $\bot_c$ rule application is the premise of the rule considered in each case. That is why the ¬-i rule was not considered, if so, the conclusion of $\bot_c$ rule would have to be a $\bot$, which is prohibited by the restriction on $\bot_c$-rule. ☐

Rule ⊑-e

$$
\cfrac{
\begin{array}{c}
[\neg\alpha] \\
\Pi_1 \\
\cfrac{\bot}{\alpha} \quad \cfrac{\Pi_2}{\alpha \sqsubseteq \beta}
\end{array}
}{
\beta
}
\qquad \triangleright \qquad
\cfrac{
\begin{array}{c}
\cfrac{[\alpha]^1 \quad \cfrac{\Pi_2}{\alpha \sqsubseteq \beta}}{\beta} \quad [\neg\beta]^2 \\
\hline
\cfrac{\bot}{\neg\alpha} \ 1 \\
\Pi_1 \\
\cfrac{\bot}{\beta} \ 2
\end{array}
}{}
$$

The reductions below will be used in the induction step in Theorem 5.

Let Π be a deduction of α from Ω which contains a maximal formula occurrence F. We say that Π' is a reduction of Π at F if we obtain Π' by removing F using the reductions below. Since F clearly can not be atomic, each reduction refers to a possible principal sign of F. If the principal sign of F is ψ, then Π' is said to be a ψ-reduction of Π. In each case, one can easily verify that Π' obtained is still a deduction of α from Ω.

⊓-reduction

$$
\frac{\dfrac{\Pi_1}{\forall L_\alpha} \quad \dfrac{\Pi_2}{\forall L_\beta}}{\dfrac{\forall L_{(\alpha \sqcap \beta)}}{\forall L_\alpha}} \qquad \triangleright \qquad \dfrac{\Pi_1}{\forall L_\alpha}
$$

∀-reduction

$$
\frac{\dfrac{\dfrac{\Pi_1}{L,\forall R_\alpha}}{L_{\forall R.\alpha}}}{L,\forall R_\alpha} \qquad \triangleright \qquad \dfrac{\Pi_1}{L,\forall R_\alpha}
$$

¬-reduction

$$
\frac{\dfrac{\dfrac{[^L\alpha]}{\Pi_1}}{\dfrac{\bot}{\neg^L\neg\alpha}} \quad \dfrac{\Pi_2}{^L\alpha}}{\bot} \qquad \triangleright \qquad \dfrac{\dfrac{\Pi_2}{[^L\alpha]}}{\dfrac{\Pi_1}{\bot}}
$$

The fact that DL has no concept internalizing ⊑ imposes quite particular features on the form of the normal proofs in ND$_{\mathcal{ALC}}$.

A ND$^-{}_{\mathcal{ALC}}$ deduction is called *normal* when it does not have maximal formula occurrences. Theorem 5 shows how we can construct a normal deduction in ND$^-{}_{\mathcal{ALC}}$.

Consider a deduction Π in ND$^-{}_{\mathcal{ALC}}$. Applying Lemma 11 we obtain a new deduction Π' without any maximal ⊑-formulas. Then we apply Lemma 12 to reduce the number of applications of $\bot_c$-rule on each branch and moving the remaining downwards to the end of each branch. Without loss of generality we can from now on consider any deduction in ND$^-{}_{\mathcal{ALC}}$ as having no maximal ⊑-formula and at most one $\bot_c$-rule application per branch, namely, the last one application in the branch.

Theorem 5 (normalization of ND$_{\mathcal{ALC}}$) *If $\Omega \vdash_{\text{ND}^-{}_{\mathcal{ALC}}} \alpha$, then there is a normal deduction in ND$^-{}_{\mathcal{ALC}}$ of α from Ω.*

Proof Let Π be a deduction in ND$^-{}_{\mathcal{ALC}}$ having the form remarked in the previous paragraph. Consider the pair (d, n) where d is the maximum degree among the maximal formulas, and n is the number of maximal formulas with degree d. We proceed the normalization proof by induction on the lexicographic pair (d, n).

Let F be one of the highest maximal formula with degree d and consider each possible case according the principal sign of F.

If F has as principal sign $\sqcap$, applying the $\sqcap$-reduction we get a new deduction Π_1 with complexity (d_1, n_1). We now have $d_1 \leq d$, depending on the existence of other maximal $\sqcap$-formulas on Π. If $d_1 = d$, then necessarily $n_1 < n$. The cases where the principal sign of F is $\neg$ or $\forall$ are similar. Two facts can be observed. First, the $\sqsubseteq$-reduction will not be used anymore, since Π does not have any remaining maximal $\sqsubseteq$-formula. Second, although the $\neg$-reduction can increase the number of maximal formulas, those maximal formulas will undoubtedly have degree less than d, so that, we indeed have $(d_1, n_1) < (d, n)$. So, by the induction hypothesis, we have that Π_1 is normalizable and so is Π for each principal sign considered. $\quad\square$

As we have already mentioned ND$_{\mathcal{ALC}}$ has no concept internalization $\sqsubseteq$. This imposes quite particular form of the normal proofs in ND$^-_{\mathcal{ALC}}$. Consider a thread in a deduction Π in ND$^-_{\mathcal{ALC}}$, such that no element of the thread is a minor premise of $\sqsubseteq$-e rule. We shall see that if Π is normal, the thread can be divided into two parts. There is one formula occurrence A in the thread such that all formula occurrences in the thread above A are premises of applications of elimination rules and all formula occurrences below A in the thread (except the last one) are premises of applications of introduction rules. Therefore, in the first part of the thread, we start from the top-most formula and decrease the complexity of that until A. In the second part of the thread we pass to more and more complex formulas. Given that, A is said thus the minimum formula in the thread. Moreover, each branch on Π has at most one application of $\perp_c$ rule as its last rule application.

Normalization is important since from it one can provide complete procedure to produce canonical proofs in $\mathcal{ALC}$. Canonical proofs are important regarding explaining theoremhood.

References

1. Bellin, G., Hyland, M., Robinson, E., Urban, C.: Categorical proof theory of classical propositional calculus. Theor. Comput. Sci. **364**(2), 146–165 (2006)
2. Berger, U., Buchholz, W., Schwichtenberg, H.: Refined program extraction form classical proofs. Ann. Pure Appl. Logic **114**(1–3), 3–25 (2002)
3. Girard, J.Y., Taylor, P.G., Lafont, Y.: Proofs and types. Cambridge University Press, New York (1993)
4. Prawitz, D.: Natural deduction: a proof-theoretical study. Ph.D. Thesis, Almqvist & Wiksell (1965)
5. Seldin, J.: Normalization and excluded middle. I. Studia Logica **48**(2), 193–217 (1989)

Chapter 6
Towards a Proof Theory for $\mathcal{ALCQI}$

Abstract In this chapter we present a Sequent Calculus and a Natural Deduction for $\mathcal{ALCQI}$ description logic. These calculi are the first step towards extensions for the previously presented systems to more expressive description logics.

Keywords $\mathcal{ALCQI}$ · Sequent calculus · Natural deduction · Proof theory · Normalization · Completeness · Soundness

6.1 Introduction

Some practical applications require a more expressive DL. For instance, if we want to formalize and reasoning over ER or UML diagrams using DL, we will need to move to $\mathcal{ALCQI}$ [2–6]. In Sect. 7.3, we present a practical use of the $\mathrm{ND}_{\mathcal{ALCQI}}$ for reasoning over an UML diagram.

6.2 $\mathcal{ALCQI}$ Introduction

$\mathcal{ALCQI}$ is an extension of $\mathcal{ALC}$ with number restrictions and inverse roles.

$$C ::= \perp \mid A \mid \neg C \mid C_1 \sqcap C_2 \mid C_1 \sqcup C_2 \mid \exists R.C \mid \forall R.C \mid \leq nR.C \mid \geq nR.C$$
$$R ::= P \mid P^-$$

where A stands for atomic concepts and R for atomic roles. Some of the above operators can be mutually defined: (i) $\perp$ for $A \sqcap \neg A$; (ii) $\top$ for $\neg\perp$; (iii) $\geq kR.C$ for $\neg(\leq k - 1R.C)$; (iv) $\leq kR.C$ for $\neg(\geq k + 1R.C)$; (v) $\exists R.C$ for $\geq 1R.C$.

An $\mathcal{ALCQI}$ theory is a finite set of inclusion assertions of the form $C_1 \sqsubseteq C_2$. The semantics of $\mathcal{ALCQI}$ constructors and theories is analogous to that of $\mathcal{ALC}$. The semantics for qualified number restrictions are presented in Sect. 2.3. The semantics of inverse roles is:

A. Rademaker, *A Proof Theory for Description Logics*,
SpringerBriefs in Computer Science, DOI: 10.1007/978-1-4471-4002-3_6,
© The Author(s) 2012

$$\frac{}{\alpha \Rightarrow \alpha} \qquad\qquad \frac{}{\bot \Rightarrow \alpha}$$

$$n \leq m \ \frac{}{\leq nR,L \, \alpha \Rightarrow \ \leq mR,L \, \alpha} \qquad\qquad n \geq m \ \frac{}{\geq nR,L \, \alpha \Rightarrow \ \geq mR,L \, \alpha}$$

Fig. 6.1 The System $\mathrm{SC}_{\mathcal{ALCQI}}$: the axioms

$$(P^-)^{\mathcal{I}} = \{(a, a') \in \Delta^{\mathcal{I}} \times \Delta^{\mathcal{I}} \mid (a', a) \in P^{\mathcal{I}}\}$$

The next sections present a sequent calculus for $\mathcal{ALCQI}$ named $\mathrm{SC}_{\mathcal{ALCQI}}$. In Sect. 6.3 we present the system and in Sect. 6.4 we prove its soundness. The proof of $\mathrm{SC}_{\mathcal{ALCQ}}$ completeness should be obtained following the same strategy used for $\mathrm{SC}_{\mathcal{ALC}}$. A version of $\mathrm{SC}_{\mathcal{ALCQ}}$ can be designed along the same basic idea used to design the $\mathrm{SC}^{[]}{}_{\mathcal{ALC}}$. Afterwards, provision of counter-example from fully expanded trees that are not proofs must be obtained.

6.3 The Sequent Calculus for $\mathcal{ALCQI}$

The $\mathrm{SC}_{\mathcal{ALCQI}}$ sequent calculus is a conservative extension of $\mathrm{SC}_{\mathcal{ALC}}$ system to deal with qualified number restriction. The syntax for labeled concepts is modified to accept upper (at-most) and lower (at-least) bounds labels:

$$LB ::= \forall R \mid \exists R \mid \leq nR \mid \geq nR$$
$$R ::= P \mid P^-$$
$$L ::= LB, L \mid \emptyset$$
$$\phi_{lc} ::= {}^L\phi_c$$

where n range over natural numbers, R over atomic role names and C over $\mathcal{ALCQI}$ concepts.

The translation of $\mathrm{SC}_{\mathcal{ALCQI}}$ labeled concept to their $\mathcal{ALCQI}$ concept counterpart is straightforward. That is, we can easily extend the definiton of the σ function presented in Sect. 3.1. For instance, ${}^{\geq nR}\alpha$ is equivalent of $\geq nR.\alpha$ and ${}^{\leq nR}\alpha$ is equivalent of $\leq nR.\alpha$. Finally, we observe that $\mathcal{ALCNI}$ is trivially obtained from $\mathcal{ALCQI}$ if we restrict qualified number restriction labels only to the $\top$ concept.

The $\mathrm{SC}_{\mathcal{ALCQI}}$ system is presented at Figs. 6.1, 6.2, 6.3 and 6.4 where L stands for list of labels. In some rules, we superscribe the list of labels with the kind of labels allowed on it. For example, in rule $\sqcap$-l, we retrict L to contain only $\forall R$ or $\geq nR$ labels. We use the notation $L^{\forall \leq}$. Moreover, for easier understanding, some provisos regarding the order relation between the number n and m are presented on the left of some rules.

$$\frac{\Delta \Rightarrow \Gamma}{\Delta, \delta \Rightarrow \Gamma} \text{ weak-l} \qquad\qquad \frac{\Delta \Rightarrow \Gamma}{\Delta \Rightarrow \Gamma, \gamma} \text{ weak-r}$$

$$\frac{\Delta, \delta, \delta \Rightarrow \Gamma}{\Delta, \delta \Rightarrow \Gamma} \text{ contraction-l} \qquad\qquad \frac{\Delta \Rightarrow \Gamma, \gamma, \gamma}{\Delta \Rightarrow \Gamma, \gamma} \text{ contraction-r}$$

$$\frac{\Delta_1, \delta_1, \delta_2, \Delta_2 \Rightarrow \Gamma}{\Delta_1, \delta_2, \delta_1, \Delta_2 \Rightarrow \Gamma} \text{ perm-l} \qquad\qquad \frac{\Delta \Rightarrow \Gamma_1, \gamma_1, \gamma_2, \Gamma_2}{\Delta \Rightarrow \Gamma_1, \gamma_2, \gamma_1, \Gamma_2} \text{ perm-r}$$

$$\frac{\Delta_1 \Rightarrow \Gamma_1, {}^L\alpha \qquad {}^L\alpha, \Delta_2 \Rightarrow \Gamma_2}{\Delta_1, \Delta_2 \Rightarrow \Gamma_1, \Gamma_2} \text{ cut}$$

Fig. 6.2 The System $SC_{\mathcal{ALCQI}}$: structural rules

$$\frac{\Delta, {}^{L^{\forall \geq}}\alpha, {}^{L^{\forall \geq}}\beta \Rightarrow \Gamma}{\Delta, {}^{L^{\forall \geq}}(\alpha \sqcap \beta) \Rightarrow \Gamma} \sqcap\text{-l} \qquad\qquad \frac{\Delta \Rightarrow \Gamma, {}^{L^{\forall \leq}}\alpha \qquad \Delta \Rightarrow \Gamma, {}^{L^{\forall \leq}}\beta}{\Delta \Rightarrow \Gamma, {}^{L^{\forall \leq}}(\alpha \sqcap \beta)} \sqcap\text{-r}$$

$$\frac{\Delta, {}^{L^{\exists \leq}}\alpha \Rightarrow \Gamma \qquad \Delta, {}^{L^{\exists \leq}}\beta \Rightarrow \Gamma}{\Delta, {}^{L^{\exists \leq}}(\alpha \sqcup \beta) \Rightarrow \Gamma} \sqcup\text{-l} \qquad\qquad \frac{\Delta \Rightarrow \Gamma, {}^{L^{\exists \geq}}\alpha, {}^{L^{\exists \geq}}\beta}{\Delta \Rightarrow \Gamma, {}^{L^{\exists \geq}}(\alpha \sqcup \beta)} \sqcup\text{-r}$$

$$\frac{\Delta \Rightarrow \Gamma, {}^{\neg L^{\forall \exists}}\alpha}{\Delta, {}^{L^{\forall \exists}}\neg\alpha \Rightarrow \Gamma} \neg\text{-l} \qquad\qquad \frac{\Delta, {}^{\neg L^{\forall \exists}}\alpha \Rightarrow \Gamma}{\Delta \Rightarrow \Gamma, {}^{L^{\forall \exists}}\neg\alpha} \neg\text{-r}$$

Fig. 6.3 The System $SC_{\mathcal{ALCQI}}$: $\sqcap$, $\sqcup$ and $\neg$ rules

The provisos of rules $\forall$-r, $\forall$-l, prom-$\exists$, prom-$\forall$, $\sqcup$-l and $\sqcup$-r are the same presented in Sect. 3.1. Moreover, we have the following additional provisos:

- Rules $\neg$-l and $\neg$-r, the list of labels L cannot have number restrictions $\leq nR$ nor $\geq nR$ for any R;
- Rule $\sqcap$-l, L cannot have $\leq nR$ nor $\exists R$ labels;
- Rule $\sqcap$-r, L cannot have $\geq nR$ nor $\exists R$ labels;
- Rule $\sqcup$-l, L cannot have $\geq nR$ nor $\forall R$ labels;
- Rule $\sqcup$-r, L cannot have $\leq nR$ nor $\forall R$ labels;
- Rule prom-$\geq$, for all ${}^L\delta \in \Delta$, L must only contain $\geq nR$ or $\forall R$ labels. For all ${}^L\gamma \in \Gamma$, L must only contain $\geq nR$ or $\exists R$ labels.

Besides the rules inherited from $SC_{\mathcal{ALC}}$ with some extra provisos, $SC_{\mathcal{ALCQI}}$ specific rules are: (1) the four rules shift-$\leq$, $\geq$-{l,r} that increase (decrease) labels upper (lower) bounds; (2) the rules $\leq \exists$-{l,r} and $\exists \leq$-{l,r} transform quantified number restricted labels into existential and the order way around.

Before present the soundness and completeness of $\mathcal{SALC}$ system, let us first present a simple example of its usage. The following proof draws the conclusion

$$\frac{\Delta, {}^{L,\forall R}\alpha \Rightarrow \Gamma}{\Delta, {}^{L}(\forall R.\alpha)L_2 \Rightarrow \Gamma} \ \forall\text{-l} \qquad \frac{\Delta \Rightarrow \Gamma, {}^{L,\forall R}\alpha}{\Delta \Rightarrow \Gamma, {}^{L}(\forall R.\alpha)} \ \forall\text{-r}$$

$$\frac{\Delta, {}^{L,\exists R}\alpha \Rightarrow \Gamma}{\Delta, {}^{L}(\exists R.\alpha) \Rightarrow \Gamma} \ \exists\text{-l} \qquad \frac{\Delta \Rightarrow \Gamma, {}^{L,\exists R}\alpha}{\Delta \Rightarrow \Gamma, {}^{L}(\exists R.\alpha)} \ \exists\text{-r}$$

$$\frac{\Delta, {}^{L,\leq nR}\alpha \Rightarrow \Gamma}{\Delta, {}^{L}\leq nR.\alpha \Rightarrow \Gamma} \ \leq\text{-l} \qquad \frac{\Delta \Rightarrow \Gamma, {}^{L,\leq nR}\alpha}{\Delta \Rightarrow \Gamma, {}^{L}\leq nR.\alpha} \ \leq\text{-r}$$

$$\frac{\Delta, {}^{L,\geq nR}\alpha \Rightarrow \Gamma}{\Delta, {}^{L}\geq nR.\alpha \Rightarrow \Gamma} \ \geq\text{-l} \qquad \frac{\Delta \Rightarrow \Gamma, {}^{L,\geq nR}\alpha}{\Delta \Rightarrow \Gamma, {}^{L}\geq nR.\alpha} \ \geq\text{-r}$$

$$n \leq m \ \frac{\Delta, {}^{\geq nR,L}\alpha \Rightarrow \Gamma}{\Delta, {}^{\geq mR,L}\alpha \Rightarrow \Gamma} \ \text{shift-}\geq\text{-l} \qquad n \geq m \ \frac{\Delta \Rightarrow {}^{\geq nR,L}\alpha, \Gamma}{\Delta \Rightarrow {}^{\geq mR,L}\alpha, \Gamma} \ \text{shift-}\geq\text{-r}$$

$$n \geq m \ \frac{\Delta, {}^{\leq nR,L}\alpha \Rightarrow \Gamma}{\Delta, {}^{\leq mR,L}\alpha \Rightarrow \Gamma} \ \text{shift-}\leq\text{-l} \qquad n \leq m \ \frac{\Delta \Rightarrow {}^{\leq nR,L}\alpha, \Gamma}{\Delta \Rightarrow {}^{\leq mR,L}\alpha, \Gamma} \ \text{shift-}\leq\text{-r}$$

$$\frac{\Delta, {}^{\geq 1R,L}\alpha \Rightarrow \Gamma}{\Delta, {}^{\exists R,L}\alpha \Rightarrow \Gamma} \ \geq \exists\text{-l} \qquad n \geq 1 \ \frac{\Delta \Rightarrow \Gamma, {}^{\geq nR,L}\alpha}{\Delta \Rightarrow \Gamma, {}^{\exists R,L}\alpha} \ \geq \exists\text{-r}$$

$$n \geq 1 \ \frac{\Delta, {}^{\exists R,L}\alpha \Rightarrow \Gamma}{\Delta, {}^{\geq nR,L}\alpha \Rightarrow \Gamma} \ \exists \geq\text{-l} \qquad \frac{\Delta \Rightarrow \Gamma, {}^{\exists R,L}\alpha}{\Delta \Rightarrow \Gamma, {}^{\geq 1R,L}\alpha} \ \exists \geq\text{-r}$$

$$\frac{\Delta, {}^{\exists R,L_1}\alpha \Rightarrow {}^{L_2}\beta, \Gamma}{\Delta, {}^{L_1}\alpha \Rightarrow {}^{\forall R^-,L_2}\beta, \Gamma} \ \exists\text{-inv} \qquad \frac{\Delta, {}^{L_1}\alpha \Rightarrow {}^{\forall R^-,L_2}\beta, \Gamma}{\Delta, {}^{\exists R,L_1}\alpha \Rightarrow {}^{L_2}\beta, \Gamma} \ \text{inv-}\exists$$

$$\frac{\Delta \Rightarrow \Gamma}{+\geq nR\Delta \Rightarrow +\geq nR\Gamma} \ \text{prom-}\geq \qquad \frac{\delta \Rightarrow \gamma}{+\leq nR\gamma \Rightarrow +\leq nR\delta} \ \text{prom-}\leq$$

$$\frac{\delta \Rightarrow \Gamma}{+\exists R\delta \Rightarrow +\exists R\Gamma} \ \text{prom-}\exists \qquad \frac{\Delta \Rightarrow \gamma}{+\forall R\Delta \Rightarrow +\forall R\gamma} \ \text{prom-}\forall$$

Fig. 6.4 The system SC$_{\mathcal{ALCQI}}$: $\forall$, $\exists$, $\leq$, $\geq$ and *inverse* rules

everyone that has at least one child male or at least one child female have a child in $\mathcal{ALCQI}$ terms.

Example 5 In the proof below, *Fem* is an abbreviation for *Female* and *child* for *hasChild*.

$$\cfrac{\cfrac{\cfrac{\cfrac{\cfrac{\cfrac{\cfrac{Fem \Rightarrow Fem}{\exists child\, Fem \Rightarrow \exists child\, Fem}}{\geq 1child\, Fem \Rightarrow \exists child\, Fem}}{\geq 1child\, Fem \Rightarrow \exists child\, Male,\, \exists child\, Fem}}{\geq 1child\, Fem \Rightarrow \exists child\, (Male \sqcup Fem)}}{\geq 1child\, Fem \Rightarrow \exists child.(Male \sqcup Fem)}}{\geq 1child.Fem \Rightarrow \exists child.(Male \sqcup Fem)} \qquad \cfrac{\cfrac{\cfrac{\cfrac{\cfrac{\cfrac{Male \Rightarrow Male}{\exists child\, Male \Rightarrow \exists child\, Male}}{\geq 1child\, Male \Rightarrow \exists child\, Male}}{\geq 1child\, Male \Rightarrow \exists child\, Male,\, \exists child\, Fem}}{\geq 1child\, Male \Rightarrow \exists child\, (Male \sqcup Fem)}}{\geq 1child\, Male \Rightarrow \exists child.(Male \sqcup Fem)}}{\geq 1child.Male \Rightarrow \exists child.(Male \sqcup Fem)}}{\geq 1child.Male \sqcup \geq 1child.Fem \Rightarrow \exists child.(Male \sqcup Fem)}$$

6.4 SC$_{\mathcal{ALCQI}}$ Soundness

Theorem 6 ($\mathcal{SALCQ}$ is sound) *Considering Ω a set of sequents, a theory presentation or a TBox, let an Ω-proof be any $\mathcal{SALCQ}$ proof in which sequents from Ω are permitted as initial sequents (in addition to the logical axioms). The soundness of $\mathcal{SALCQ}$ states that if a sequent $\Delta \Rightarrow \Gamma$ has an Ω-proof, then $\Delta \Rightarrow \Gamma$ is satisfied by every interpretation which satisfies Ω. That is,*

$$if \quad \Omega \vdash_{SC_{\mathcal{ALCQI}}} \Delta \Rightarrow \Gamma \quad then \quad \Omega \models \bigsqcap_{\delta \in \Delta} \sigma(\delta) \sqsubseteq \bigsqcup_{\gamma \in \Gamma} \sigma(\gamma)$$

for all interpretation $\mathcal{I}$.

Proof We proof Theorem 6 by induction on the length of the Ω-proofs. The length of a Ω-proof is the number of applications for any derivation rule of the calculus.

For the base case, proofs with length zero are proofs $\Omega \vdash \Delta \Rightarrow \Gamma$ where $\Delta \Rightarrow \Gamma$ occurs in Ω. In that case, it is easy to see that the theorem holds.

As inductive hypothesis, we will consider that for proofs of length n the theorem holds. It is now sufficient to show that each of the derivation rules preserves the truth. That is, if the premises holds, the conclusion must also hold. Remembering from Sect. 3.1 that the natural interpretation of a sequent $\Delta \Rightarrow \Gamma$ (Δ and Γ range over labeled concepts) is the $\mathcal{ALC}$ formula

$$\bigsqcap_{\delta \in \Delta} \sigma(\delta) \sqsubseteq \bigsqcup_{\gamma \in \Gamma} \sigma(\gamma)$$

For a clear presentation, we will sometimes omit the translation from labeled concepts to $\mathcal{ALCQ}$ concepts and directly take Δ as the conjunction of $\mathcal{ALCQ}$ concepts and Γ as the disjunction of $\mathcal{ALCQ}$ concepts and assume that $\Delta \Rightarrow \Gamma$ has $\Delta \sqsubseteq \Gamma$ as a natural interpretation.

For rules on Fig. 6.2, we can apply standard set theory. The proof of their soundness are the same presented in Sect. 3.2 for $\mathcal{SALC}$. For instance, let us consider A, B, C, D and X sets. Rules weak-l and weak-r following from $(A \cap B) \subseteq A$ and $A \subseteq (A \cup B)$. Rules contraction-l and contraction-r follow from $A \cap A = A$ and $A \cup A = A$.

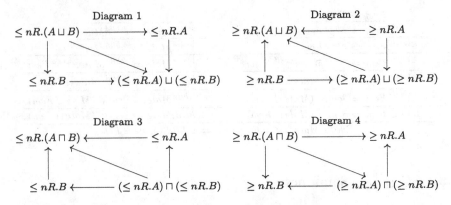

Fig. 6.5 The inclusion diagrams for $\leq$ and $\geq$ over $\sqcup$ and $\sqcap$. The arrow $A \to B$ means $A \sqsubseteq B$

In rules perm-l and perm-r, the premises and conclusions have the same semantics. The cut rule is also easily justified by set theory: if $A \subseteq (B \cup X)$ and $(X \cap C) \subseteq D$, we must have $(A \cap C) \subseteq (B \cup D)$.

In Fig. 6.4, rules $\forall$-l, $\forall$-r, $\exists$-l, $\exists$-r, $\leq$-l, $\leq$-r, $\geq$-l and $\geq$-r represent steps in the translation of labeled concepts to $\mathcal{ALCQ}$ concepts (reading top-bottom), so that, premises and conclusion have the same semantics, if the former subsumption holds, the later will also hold.

Rule $\exists \geq$-l is sound regarding the $\mathcal{SALCQ}$ semantic fact that $\geq nR.A \sqsubseteq \exists R.A$ if $n \geq 1$. If we take $A = \Delta^{\mathcal{I}}$, $B = \Gamma^{\mathcal{I}}$, $C = (^{\geq 1R,L}\alpha)^{\mathcal{I}}$ and $D = (^{\exists R,L}\alpha)^{\mathcal{I}}$ for any given $\mathcal{I}$. Then we can conclude that if $A \cap C \subset B$ (premise) and $C \subset D$ (fact) then $A \cap D \subset B$ (conclusion).

The argument to show rule $\exists \geq$-r soundness is similar, Considering now the fact that $\exists R.A \equiv \geq 1R.A$ follows from the $\mathcal{ALCQ}$ semantics, we can show that: if we take $A = \Delta^{\mathcal{I}}$, $B = \Gamma^{\mathcal{I}}$, $C = (^{\exists R,L}\alpha)^{\mathcal{I}}$ and $D = (^{\geq 1R,L}\alpha)^{\mathcal{I}}$ for any given $\mathcal{I}$, then if $A \subset B \cup C$ (premise) and $C \equiv D$ (fact) then $A \subset B \cup D$ (conclusion).

Rules $\neg$-l and $\neg$-r do not deal with quantified labeled concepts, their soundess were provided in Sect. 3.2.

From the $\mathcal{ALCQ}$ semantics, we know that if $n \leq m$: (1) $\geq mR.C \sqsubseteq \geq nR.C$; and (2) $\leq nR.C \sqsubseteq \leq mR.C$ for any concept C. Taking $A = \Delta^{\mathcal{I}}$ and $B = \Gamma^{\mathcal{I}}$ for any $\mathcal{I}$, rules shift-$\geq$-l and shift-$\leq$-r are sound:

- if $A \cap (^{\geq nR,L}\alpha)^{\mathcal{I}} \subseteq B$ (premise), and $^{\geq mR,L}\alpha \subseteq {}^{\geq nR,L}\alpha$ (by 1 if $n \leq m$), then $A \cap (^{\geq mR,L}\alpha)^{\mathcal{I}} \subseteq B$ (conclusion);
- if $A \subseteq (^{\leq nR,L}\alpha)^{\mathcal{I}} \cup B$ (premise) and $^{\leq nR,L}\alpha \subseteq {}^{\leq mR,L}\alpha$ (by 2 if $n \leq m$), then $A \subseteq (^{\leq mR,L}\alpha)^{\mathcal{I}} \cup B$ (conclusion);

Rules shift-$\leq$-l and shift-$\geq$-r are similar, using the same semantics facts 1 and 2 above.

For rules $\sqcup$-l, $\sqcup$-r, prom-$\exists$, prom-$\forall$, $\sqcap$-l and $\sqcap$-r we use the inclusion relations expressed in the diagrams of Fig. 6.5. The arrows in the Figure indicate the inclusion direction, that is, if $A \to B$, than $A \sqsubseteq B$. Following the traditional proof theory terminology for sequent calculi, we call the principal formula, the formula occurring in the lower sequent of the inference which is not in the designated sets (Δ and Γ) and the auxiliary formulas are the formulas from the premises, subformulas of the principal formula in the conclusion.

Rule $\sqcup$-l with the proviso that the lists labels in auxiliary formulas can only contain $\exists R$ or $\leq nR$ labels for any role R and integer n is sound. This follows from: (1) the diagram 1 in the figure that shows that the union of the interpretation of auxiliary formulas is subset of the interpretation of the principal formula; and (2) the set theory fact that if $A \subseteq C$, $B \subseteq C$ and $X \subseteq A \cup B$ then $X \subseteq C$.

Rule $\sqcup$-r with the proviso that the list of labels in auxiliary formulas does not contain labels rather than $\exists R$ and $\geq nR$ for any role R and integer n is also sound. This follows from: (1) diagram 2 which shows that the interpretation of the principal formula contains the union of the interpretation of the auxiliary formulas; and (2) the set theory fact that if $A \subseteq B \cup C$ and $B \cup C \subseteq X$ then $A \subseteq X$.

Rule $\sqcap$-l providing that labels of auxiliary formulas does not contain labels rather than $\forall R$ and $\geq nR$ is sound given that: (1) diagram 4 shows that the intersection of the (interpretation of) the premises contains the interpretation of the conclusion, for any interpretation function; and (2) the set theory transitive property of the inclusion relation, that is, if $A \cap B \subseteq C$ and $X \subseteq A \cap B$ then $X \subseteq C$.

The soundness of rule $\sqcap$-r, providing that the list of labels of auxiliary formulas contain only $\forall$ and $\leq nR$ labels is proved with: (1) diagram 3 that shows that the intersection of the interpretation of the auxiliary formulas is included in the principal formula; (2) the fact that if $A \subseteq B$, $A \subseteq C$ and $B \cap C \subseteq X$ then $A \subseteq X$.

The proof of rules inv-$\exists$ and $\exists$-inv soundness derives from the fact that $A \sqsubseteq \forall R^-.B$ if and only if $\exists R.A \sqsubseteq B$. For clear presentation, we can state this fact as a rule in a natural deduction style:

$$\frac{(2) \quad \exists R.A \sqsubseteq B}{(1) \quad A \sqsubseteq \forall R^-.B} \; inv*$$

Now we only have to prove the double soundess of the above rule and consider $A \equiv {}^{L_1}\alpha$ and $B \equiv {}^{L_2}\beta$.

Case 1 → 2. Let $v \in \exists R.A^{\mathcal{I}} = \{v \mid (v, u) \in R^{\mathcal{I}} \wedge u \in A^{\mathcal{I}}\}$ thus $\exists u \in A^{\mathcal{I}}$ such that $(v, u) \in R^{\mathcal{I}}$ and hence $(u, v) \in (R^-)^{\mathcal{I}}$. But from (1) we have that $u \in \forall R^-.B^{\mathcal{I}}$, thus $\forall v((u, v) \in (R^-)^{\mathcal{I}} \rightarrow v \in B^{\mathcal{I}})$, hence $v \in B^{\mathcal{I}}$ we conclude (2). Note also that this conclusion also holds if $R^{\mathcal{I}} = \emptyset$.

Case 2 → 1. Let us assume that there is a $(v, u) \in R^{\mathcal{I}}$, so, $v \in \exists R.A^{\mathcal{I}}$ and hence $v \in B^{\mathcal{I}}$, by (2). We have $(u, v) \in (R^-)^{\mathcal{I}}$ so $\forall v((u, v) \in (R^-)^{\mathcal{I}} \rightarrow v \in B^{\mathcal{I}})$ and hence $u \in \forall R^-.B^{\mathcal{I}}$. If for some $u \in A^{\mathcal{I}}$ there is no v such that $(v, u) \in R^{\mathcal{I}}$ then $u \in \forall R^-.B^{\mathcal{I}}$, vacuously. $\square$

6.5 On SC$_{\mathcal{ALCQI}}$ Completeness

The proof of SC$_{\mathcal{ALCQI}}$ completeness should be obtained following the same strategy used for SC$_{\mathcal{ALC}}$. A deterministic version of SC$_{\mathcal{ALCQI}}$ can be designed along the same basic idea used on SC$^{[]}_{\mathcal{ALC}}$. Afterwards, provision of counter-example from fully expanded trees that are not proofs must be obtained.

Next, we briefly show how to provide a counter-example for a top-sequent that is not an axiom (initial sequent) in a fully expanded tree. Let us consider the full expanded tree in the sequel.

Example 6 The bottom sequent represents an unsatisfiable subsumption. Clearly, it is not true that all people with at least two children necessarily have one child male and the other female. In the proof, F stands for *Female*, M for *Male* and *child* for *hasChild*.

$$
\dfrac{
\dfrac{
\dfrac{
\dfrac{
\dfrac{
\dfrac{\dfrac{M \Rightarrow M}{{}^{\exists child}M \Rightarrow {}^{\exists child}M} \qquad \dfrac{F \Rightarrow M}{{}^{\exists child}F \Rightarrow {}^{\exists child}M}}{{}^{\exists child}(M \sqcup F) \Rightarrow {}^{\exists child}M}
}{{}^{\geq 1 child}(M \sqcup F) \Rightarrow {}^{\exists child}M}
}{{}^{\geq 2 child}(M \sqcup F) \Rightarrow {}^{\exists child}M}
}{{}^{\geq 2 child}(M \sqcup F) \Rightarrow \exists child.M} \qquad
\dfrac{
\dfrac{
\dfrac{
\dfrac{\dfrac{M \Rightarrow F}{{}^{\exists child}M \Rightarrow {}^{\exists child}F} \qquad \dfrac{F \Rightarrow F}{{}^{\exists child}F \Rightarrow {}^{\exists child}F}}{{}^{\exists child}(M \sqcup F) \Rightarrow {}^{\exists child}F}
}{{}^{\geq 1 child}(M \sqcup F) \Rightarrow {}^{\exists child}F}
}{{}^{\geq 2 child}(M \sqcup F) \Rightarrow {}^{\exists child}F}
}{{}^{\geq 2 child}(M \sqcup F) \Rightarrow \exists child.F}
}{{}^{\geq 2 child}(M \sqcup F) \Rightarrow \exists child.M \sqcap \exists child.F}
}{\geq 2 child.(M \sqcup F) \Rightarrow \exists child.M \sqcap \exists child.F}
$$

Starting from any top-sequent that is not initial, one can easily construct an interpretation $\mathcal{I}$ such that

$$\mathcal{I} \not\models\; \geq 2 hasChild.(Male \sqcup Female) \sqsubseteq \exists hasChild.Male \sqcap \exists hasChild.Female$$

Following [1, section 2.3.2.1] style, we use ABox assertions to represent the restrictions about the interpretation $\mathcal{I} = (\Delta, \text{i})$ that we intend to construct. We started from the top-sequent *Female* $\Rightarrow$ *Male* and constructed $\mathcal{A}_1$ that falsifies it. The ABox $\mathcal{A}_2$, an extension of $\mathcal{A}_1$, is then constructed to falsify ${}^{\exists hasChild}Female \Rightarrow {}^{\exists hasChild}Male$. $\mathcal{A}_2$ falsifies all subsequent sequents until

$$\geq n\, hasChild\, (Male \sqcup Female) \Rightarrow {}^{\exists hasChild}Male$$

is reached. In order to falsify it we constructed $\mathcal{A}_3$ from $\mathcal{A}_2$. The main idea is that for each rule application, given a interpretation that falsifies its premise, one can provide an interpretation that falsifies its conclusion. From the natural interpretation of a sequent, Sect. 3.1 we know that in order to falsify a sequent $\Delta \Rightarrow \Gamma$, an interpretation must contain an element c such that $c \in \Delta^{\mathcal{I}}$ and $c \notin \Gamma^{\mathcal{I}}$.

$$
\begin{aligned}
\mathcal{A}_1 &= \{Female(f_1)\} \\
\mathcal{A}_2 &= \mathcal{A}_1 \cup \{hasChild(a, f_1)\} \\
\mathcal{A}_3 &= \mathcal{A}_2 \cup \{hasChild(a, f_2), Female(f_2)\}
\end{aligned}
\qquad (6.1)
$$

The desired interpretation $\mathcal{I}$ can then be extracted from $\mathcal{A}_3$:

$$\Delta^{\mathcal{I}} = \{a, f_1, f_2\}, \; \mathit{Female}^{\mathcal{I}} = \{f_1, f_2\}, \; \mathit{hasChild}^{\mathcal{I}} = \{(a, f_1), (a, f_2)\} \quad (6.2)$$

6.6 A Natural Deduction for $\mathcal{ALCQI}$

The Natural Deduction for $\mathcal{ALCQI}$, named ND$_{\mathcal{ALCQI}}$, is presented in Fig. 6.6. ND$_{\mathcal{ALCQI}}$ is an extension of the system ND$_{\mathcal{ALC}}$ presented in Chap. 5.

When dealing with theories, sometimes is more convenient to have the following rule, since theories must be closed under generalizations.

$$\frac{\alpha \sqsubseteq \beta}{\forall R.\alpha \sqsubseteq \forall R.\alpha}$$

ND$_{\mathcal{ALCQI}}$ normalization and completeness is not presented here. A completeness proof for ND$_{\mathcal{ALCQI}}$ should follow from a (technically heavy) mapping from a complete Sequent Calculus for $\mathcal{ALCQI}$ to ND$_{\mathcal{ALCQI}}$.

Assuming that normalization holds for ND$_{\mathcal{ALCQI}}$, one can define a proof procedure for ND$_{\mathcal{ALCQI}}$. Initially decompose the (candidate) conclusion ($\alpha \sqsubseteq \beta$) by means of introduction rules applied bottom-up, until atomic labeled concepts. For each atomic concept, one chooses a hypothesis from Δ and by decomposing it, by means of elimination rules, tries to achieve this very atomic (labeled) concept. This allows us to derive a (complete) proof procedure for the logic, decomposing the conclusions and the hypothesis until atomic levels and proving one set using the other. In our case we are interested in applying this proof procedure on top of theories. In the sequel we show ND$_{\mathcal{ALCQI}}$ soundness.

6.7 ND$_{\mathcal{ALCQI}}$ Soundness

This section extends the results of Sect. 5.3 to prove that ND$_{\mathcal{ALCQI}}$ rules are sound. We adopted here the same notations used in Sect. 5.3. Moreover, most part of the proof use results from Sect. 6.4.

Theorem 7 ND$_{\mathcal{ALCQI}}$ *is sound regarding the standard semantics of* $\mathcal{ALCQI}$. *That is,*

$$\text{if} \;\; \Omega \vdash \gamma \;\; \text{then} \;\; \Omega \models \gamma$$

Proof It follows direct from Lemma 13.   □

Lemma 13 *Let* Π *be a deduction in* ND$_{\mathcal{ALCQI}}$ *of F with all hypothesis in* $\Omega = (\mathcal{C}, \mathcal{S})$, *then if F is a concept:*

$$\frac{L^{\vee\geq}(\alpha\sqcap\beta)}{L^{\vee\geq}\alpha}\ \sqcap\text{-e} \qquad \frac{L^{\vee\geq}(\alpha\sqcap\beta)}{L^{\vee\geq}\beta}\ \sqcap\text{-e} \qquad \frac{L^{\vee\leq}\alpha \quad L^{\vee\leq}\beta}{L^{\vee\leq}(\alpha\sqcap\beta)}\ \sqcap\text{-i}$$

$$\frac{L^{\exists\leq}(\alpha\sqcup\beta) \quad \overset{[L^{\exists\leq}\alpha]}{\overset{\vdots}{\gamma}} \quad \overset{[L^{\exists\leq}\beta]}{\overset{\vdots}{\gamma}}}{\gamma}\ \sqcup\text{-e} \qquad \frac{L^{\exists\geq}\alpha}{L^{\exists\geq}(\alpha\sqcup\beta)}\ \sqcup\text{-i} \qquad \frac{L^{\exists\geq}\beta}{L^{\exists\geq}(\alpha\sqcup\beta)}\ \sqcup\text{-i}$$

$$\frac{L^{\vee\exists}\alpha \quad \neg L^{\vee\exists}\neg\alpha}{\bot}\ \neg\text{-e} \qquad \frac{\overset{[L^{\vee\exists}\alpha]}{\overset{\vdots}{\bot}}}{\neg L^{\vee\exists}\neg\alpha}\ \neg\text{-i} \qquad \frac{\overset{[\neg L^{\vee\exists}\neg\alpha]}{\overset{\vdots}{\bot}}}{L^{\vee\exists}\alpha}\ \bot_c$$

$$\frac{L^{\exists}R.\alpha}{L,\exists R.\alpha}\ \exists\text{-e} \qquad \frac{L,\exists R.\alpha}{L^{\exists}R.\alpha}\ \exists\text{-i} \qquad \frac{L^{\forall}R.\alpha}{L,\forall R.\alpha}\ \forall\text{-e}$$

$$\frac{L,\forall R.\alpha}{L^{\forall}R.\alpha}\ \forall\text{-i} \qquad \frac{L^{\leq}nR.\alpha}{L,\leq nR.\alpha}\ \leq\text{-e} \qquad \frac{L,\leq nR.\alpha}{L^{\leq}nR.\alpha}\ \leq\text{-i}$$

$$\frac{L^{\geq}nR.\alpha}{L,\geq nR.\alpha}\ \geq\text{-e} \qquad \frac{L,\geq nR.\alpha}{L^{\geq}nR.\alpha}\ \geq\text{-i}$$

$$\frac{\exists R,L_\alpha}{\geq 1R,L_\alpha}\ \geq\exists \qquad \frac{\geq nR,L_\alpha}{\exists R,L_\alpha}\ \exists\geq (n\geq 1)$$

$$\frac{\geq mR,L_\alpha}{\geq nR,L_\alpha}\ -\geq (m\geq n) \qquad \frac{\leq mR,L_\alpha}{\leq nR,L_\alpha}\ +\geq (m\leq n) \qquad \frac{L_\alpha}{\forall R,L_\alpha}\ Gen$$

$$\frac{L_1\alpha \quad L_1\alpha\sqsubseteq L_2\beta}{L_2\beta}\ \sqsubseteq\text{-e} \qquad \frac{\overset{[L_1\alpha]}{\overset{\vdots}{L_2\beta}}}{L_1\alpha\sqsubseteq L_2\beta}\ \sqsubseteq\text{-i} \qquad \frac{\exists R,L_1\alpha\sqsubseteq L_2\beta}{L_1\alpha\sqsubseteq\forall R^-,L_2\beta}\ inv$$

Fig. 6.6 The Natural Deduction system for $\mathcal{ALCQI}$

$$\mathcal{S}\models\bigsqcap_{A\in C}A\sqsubseteq F$$

and if F is a subsumption $A_1\sqsubseteq A_2$:

$$\mathcal{S}\models\bigsqcap_{A\in C}A\sqcap A_1\sqsubseteq A_2$$

Proof The proof of Lemma 10 is done by induction on the height of a proof tree Π represented by $|\Pi|$. The proof of ND$_{\mathcal{ALCQI}}$ rules soundness is similar to the proof of soundness of their counterparts in ND$_{\mathcal{ALC}}$.

Base case. This case is similar to the proof of Lemma 13. If $\mid \Pi \mid = 1$ then $\Omega \vdash {}^{L}\alpha$ is such that ${}^{L}\alpha$ is in Ω. In that case, is easy to see that Lemma 13 holds since by basic set theory $(A \cap B) \subseteq A$ for all A and B.

Rule $\sqcap$-e. this rule has one additional proviso that must be taken into account, namely, besides $\forall R$ roles, the label of the premise may only contain $\geq nR$ roles.

$$\Pi_1$$

By induction hypothesis, if ${}^{L}(\alpha \sqcap \beta)$ is a derivation with all hypothesis in $\{\mathcal{C}, \mathcal{S}\}$ then $\mathcal{S} \models \mathcal{C} \sqsubseteq {}^{L}(\alpha \sqcap \beta)$. From Diagram 4 on Fig. 6.5 and Axiom 2.1 we know that ${}^{L}(\alpha \sqcap \beta) \sqsubseteq {}^{L}\alpha \sqcap {}^{L}\beta$ and from basic set theory ${}^{L}\alpha \sqcap {}^{L}\beta \sqsubseteq {}^{L}\alpha$ so $\mathcal{S} \models \mathcal{C} \sqsubseteq {}^{L}\alpha$ as desired.

Rule $\sqcap$-e. let us take the proof of soundness of its counterpart in Sect. 5.3 and consider the additional proviso that L may only contain $\forall R$ and $\leq nR$ labels. Given $\mathcal{S}_1 \cup \mathcal{S}_2 \models (\mathcal{C}_1 \sqcap \mathcal{C}_2) \sqsubseteq {}^{L}\alpha \sqcap {}^{L}\beta$ (by arguments of Sect. 5.3) and ${}^{L}\alpha \sqcap {}^{L}\beta \sqsubseteq {}^{L}(\alpha \sqcap \beta)$ by Diagram 3 on Fig. 6.5 and Axiom 2.1, we can write $\mathcal{S}_1 \cup \mathcal{S}_2 \models (\mathcal{C}_1 \sqcap \mathcal{C}_2) \sqsubseteq {}^{L}(\alpha \sqcap \beta)$.

Rules $\sqcup$-e and $\sqcup$-i. As in the cases above, the proof is similar to their counterparts in Sect. 5.3. We also have to consider diagrams 1 and 2 on Fig. 6.5 to prove that $L^{\exists \geq}\alpha \sqcup L^{\exists \geq}\beta \sqsubseteq L^{\exists \geq}(\alpha \sqcup \beta)$ and $L^{\exists \leq}(\alpha \sqcup \beta) \sqsubseteq L^{\exists \leq}\alpha \sqcup L^{\exists \leq}\beta$.

Rules $\neg$-{i,e} and $\perp$-c are the same of ND$_{\mathcal{ALC}}$ since they do not handle number restrictions and inverse. Rules $\forall$-{i,e}, $\exists$-{i,e}, $\leq$-{i,e} and $\geq$-{i,e} have the same semantics of their premise and conclusion, thus they are sound.

The soundness of $- \geq$ and $+ \geq$ are direct consequence of the $\mathcal{ALCQI}$ semantics and they are actually used to prove the soundness of SC$_{\mathcal{ALCQI}}$ shift rules in Sect. 6.4. Rule inv is not only sound but also double sound, once more, we point to the proof of soundness in Sect. 6.4.

The soundness of the remain rules Gen and $\sqsubseteq$-{i,e} are consequence of the soundness of their counterparts in ND$_{\mathcal{ALC}}$, see Sect. 5.3. $\qquad \square$

References

1. Baader, F.: The Description Logic Handbook: Theory, Implementation, and Applications. Cambridge University Press, Cambridge (2003)
2. Berardi, D., Calvanese, D., De Giacomo, G.: Reasoning on UML class diagrams. Artif. Intell. **168**(1–2), 70–118 (2005)
3. Calvanese, D., De Giacomo, G., Lembo, D., Lenzerini, M., Rosati, R.: Conceptual modeling for data integration. In: Borgida, A., Chaudhri, V., Giorgini, P., Yu, E. (eds.) John Mylopoulos Festschrift, Lecture Notes in Computer Science, vol. 5600. Springer (2009). to appear
4. Calvanese, D., De Giacomo, G., Lenzerini, M., Nardi, D., Rosati, R.: Information integration: Conceptual modeling and reasoning support. In: Proceedings of the 6th International Conference on Cooperative, Information Systems (CoopIS'98), pp. 280–291 (1998)
5. Calvanese, D., Giacomo, G.D., Lenzerini, M., Rosati, R., Vetere, G.: DL-Lite: Practical reasoning for rich DLs. In: Proceedings of the 2004 Description Logic Workshop, DL 2004, CEUR Electronic Workshop Proceedings, vol. 104 (2004). http://ceur-ws.org
6. Calvanese, D., Lenzerini, M., Nardi, D.: Description logics for conceptual data modeling. In: Chomicki, J., Saake, G. (eds.) Logics for Databases and Information Systems, pp. 229–263. Kluwer Academic Publisher, Dordrecht (1998)

Chapter 7
Proofs and Explanations

Abstract In this chapter, we present some motivation for proofs explanations. For the tasks of providing proofs and explanations, we compare three deduction systems. In the sequel, we present an example of proof and possible explanation of this same proof in our system $SC_{\mathcal{ALC}}$. Finally, in the last section of this chapter, we use examples of DL deductions from Berardi et al. (Artif Intell **168**:84, 2005), using $ND_{\mathcal{ALCQI}}$ to reason on the $\mathcal{ALCQI}$ KB. The idea is to exemplify how one can obtain from $ND_{\mathcal{ALCQI}}$ proofs, a more precise and direct explanation.

Keywords Proof explanation · Proof · Deduction · Counter-example · Model · Satisfiability · Model checking

7.1 Introduction

From a logical point of view, the conceptual modeling tasks can be summarized by the following steps:

1. Observe the "world";
2. Determine what is relevant;
3. Choose or define your terminology by means of non-logical linguistic terms;
4. Write down the main laws, the axioms, governing your "world";
5. Verify the correctness (sometimes completeness too) of your set of laws, that is, the theory constructed.

Steps 1, 2 and 3 may be facilitated by the use of an informal notation (UML, ER, Flow-Charts, etc) and their respective methodology, but it is essentially "Black Art" [3]. Step 4 demands quite a lot of knowledge of the "world" begin specified (the model). Step 5 essentially provides finitely many tests as support for the correctness of an infinitely quantified property.

A deduction of a proposition α from a set of hypothesis Γ is essentially a mean of convincing that Γ entails α. When validating a theory, represented by a set of logical

A. Rademaker, *A Proof Theory for Description Logics*,
SpringerBriefs in Computer Science, DOI: 10.1007/978-1-4471-4002-3_7,

formulas, we mainly test entailments, possibly using a theorem prover. Considering a model M specified by the set of axioms $Spec(M)$, given a property ϕ about M, from the entailment tests results one can rise the following questions:

1. If $M \models \phi$ and $Spec(M) \vdash \phi$, why ϕ is truth? One must provide a proof of ϕ;
2. If $M \models \phi$, but $Spec(M) \nvdash \phi$ from the attempt to construct the proof of ϕ one may obtain a counter-model and from that counter-model an explanation for the failed entailment. Model-checking based reasoning can be used in such situation;
3. If $M \nvDash \phi$, but $Spec(M) \vdash \phi$, why does this false proposition hold? In this case, one must provide a proof of ϕ.

Here we are interested in the last case, tests providing a false positive answer, that is, the prover shows a deduction/proof for an assertion that must be invalid in the theory considered. This is one of the main reasons to explain a theorem when validating a theory. We need to provide explanation on why a false positive is entailed. Another reason to provide explanations of theorem has to do with providing explanation on why some assertion is a true positive, which is the first case. This latter use is concerned with certification; in this case the proof/deduction itself serves as a certification document. This section does not take into account educational uses of theorem provers, and their resulting theorems, since explanations in these cases are more demanding.

For the tasks of providing proofs and explanations, we compare three deduction systems, Analytic Tableaux (**AT**) [5], Sequent Calculus (**SC**) [6] and Natural Deduction (**ND**) [4] as presented in the respective references. In this section we consider the propositional logic (Minimal, Intuitionistic and Classical, as defined in [4]). Let us consider a theory (presented by a knowledge base $\mathcal{KB}$) containing the single axiom

$$\mathcal{KB} \equiv (Quad \wedge PissOnFireHydrant) \rightarrow Dog$$

which classifies a *dog* as a *quadruped* which pisses on a fire hydrant. This $\mathcal{KB}$ draws the following proposition

$$(Quad \rightarrow Dog) \vee (PissOnFireHydrant \rightarrow Dog)$$

Figure 7.1 presents three from many more possible proofs of this entailment in Propositional Tableaux system. Figure 7.2 presents three possible proofs in Sequent Calculus, they are also sorted out from many other possible proofs in Sequent Calculus. Figure 7.3 presents the only two possible normal proofs for this entailment.

Consider the derivations from Figs. 7.1 and 7.2. They correspond to the Natural Deduction derivations that are showed in Fig. 7.3. The Tableaux and Sequent Calculus variants only differ in the order of rule applications. In **ND** there is no such distinction. In this example, the order of application is irrelevant in terms of explanation, although it is not for the prover's implementation. The pattern represented by the **ND** deduction is close to what one expects from an argument drawing a conclusion from any conjunction that it contains. This example shows how **SC** proofs carry

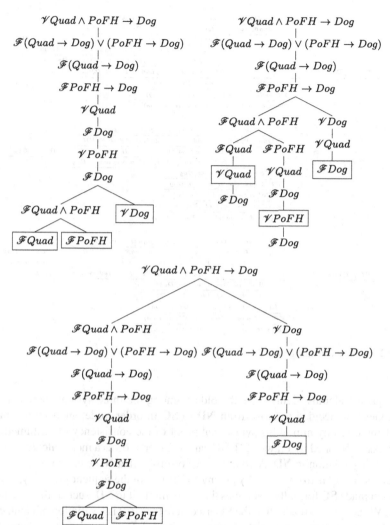

Fig. 7.1 Tableaux proofs

more information than that needed for a meaningful explanation. Concerning the **AT** system, Smullyan [5] noted that **AT** proofs correspond to **SC** proofs by considering sequents formed by positively signed formulas ($T\alpha$) at the antecedent and negatively signed ones ($F\alpha$) appearing at the succedent. A Block **AT** is defined then by considering **AT** expansion rules in the form of inference rules. In this way, our example in **SC** would carry the same content useful for explanation carried by the **AT** proofs. We must note that the different **SC** proofs and its corresponding **AT** proofs, as the ones shown, are represented, all of them, by only two possible variations of normal derivations in **ND**.

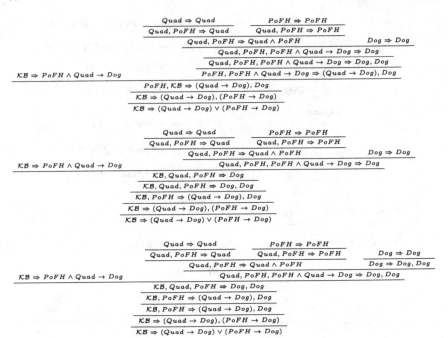

Fig. 7.2 Sequent calculus proofs

Sequent Calculus seems to be the oldest among the three systems here considered. Gentzen decided to move from **ND** to **SC** in order to detour from technical problems faced by him in his syntactical proof of the consistency of Arithmetic in 1936. As mentioned by Prawitz [4], **SC** can be understood as a meta-calculus for the deducibility relation in **ND**. A consequence of this is that **ND** can represent in only one deduction of α from $\gamma_1, \ldots, \gamma_n$ many **SC** proofs of the sequent $\gamma_1, \ldots, \gamma_n \Rightarrow \alpha$. Gentzen made **SC** formally state rules that were implicit in **ND**, such as the structural rules. We advice the reader that the **SC** used here (see [6]) is a variation of Gentzen's calculus designed with the goal of having, in each inference rule, any formula occurring in a premise as a sub-formula of some formula occurring in the conclusion. This sub-formula property facilitates the implementation of a prover based on this very system.

Consider a normal **ND** deduction Π_1 of α from $\gamma_1, \ldots, \gamma_k$, and, a deduction Π_2 of γ_i (for some $i = 1, k$) from $\delta_1, \ldots, \delta_n$. Using latter Π_1 in the former Π_2 deduction yields a (possibly non-normal) deduction of α from $\gamma_1, \delta_1, \ldots, \gamma_k, \delta_n$. This can be done in **SC** by applying a cut rule between the proofs of the corresponding sequents $\delta_1, \ldots, \delta_n \Rightarrow \gamma_i$ and $\gamma_1, \ldots, \gamma_k \Rightarrow \alpha$ yielding a proof of the sequent $\gamma_1, \delta_1, \ldots, \gamma_k, \delta_n \Rightarrow \alpha$. The new **ND** deduction can be normalized, in the former case, and the cut introduced in the latter case can be eliminated. In the case of **AT**, the fact that they are closed by *modus ponens* implies that closed **AT** for $\delta \to \gamma$ and $\gamma \to \alpha$ entails the existence of a closed **AT** for $\delta \to \alpha$. The use of cuts, or

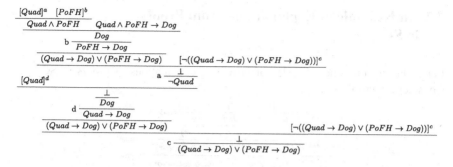

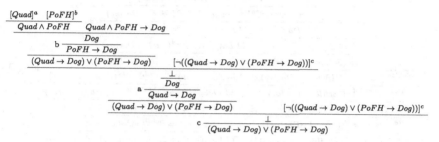

Fig. 7.3 Natural deduction proofs

equivalently, lemmas may reduce the size of a derivation. However, the relevant information conveyed by a deduction or proof in any of these systems has to firstly consider normal deductions, cut-free proofs and analytic Tableaux. They are the most representative formal objects in each of these systems as a consequence of the sub-formula property, holding in **ND** too. Besides that they are computationally easier to build than their non-normal counterparts.

These examples are carried out in Minimal Logic. For Classical reasoning, an inherent feature of most DLs, including $\mathcal{ALC}$, the above scenario changes. Any classical proof of the sequent $\gamma_1, \gamma_2 \Rightarrow \alpha_1, \alpha_2$ corresponds a **ND** deduction of $\alpha_1 \vee \alpha_2$ from γ_1, γ_2, or, of α_1 from $\gamma_1, \gamma_2, \neg\alpha_2$, or, of α_2 from $\gamma_1, \gamma_2, \neg\alpha_1$, or, of $\neg\gamma_1$ from $\neg\alpha_1, \gamma_2, \neg\alpha_2$, and so on. In Classical logic,[1] each **SC** may represent more than one deduction, since we have to choose which formula will be the conclusion in the **ND** side. We recall that it still holds that to each **ND** deduction there is more than one **SC** proof. In order to serve as a good basis for explanations of classical theorems we choose **ND** as the most adequate. Note that we are not advocating that the prover has to produce **ND** proofs directly. An effective translation to a **ND** might be provided. Of course there must be a **ND** for the logic involved. If, besides that, a normalization is provided for a system, we know that it is possible to always deal with canonical proofs satisfying the sub-formula principle.

---

[1] Intuitionistic Logic and Minimal Logic have similar behavior concerning the relationship between their respective systems of ND and SC.

7.2 An Example of Explanations From Proofs in SC$_{\mathcal{ALC}}$

Let us briefly introduce the idea of providing explanations of proofs in SC$_{\mathcal{ALC}}$. Consider the proof:

$$
\cfrac{
  \cfrac{
    \cfrac{
      \cfrac{
        \cfrac{
          \cfrac{
            \cfrac{
              \cfrac{
                \cfrac{
                  \cfrac{Doctor \Rightarrow Doctor}{Doctor \Rightarrow Rich, Doctor}\;\text{weak-r}
                }{Doctor \Rightarrow (Rich \sqcup Doctor)}\;\text{⊔-r}
              }{^{\forall child}Doctor \Rightarrow\; ^{\forall child}(Rich \sqcup Doctor)}\;\text{prom-}\forall
            }{\top, ^{\forall child}Doctor \Rightarrow\; ^{\forall child}(Rich \sqcup Doctor)}\;\text{weak-l}
          }{\top \Rightarrow\; ^{\exists child}\neg Doctor, ^{\forall child}(Rich \sqcup Doctor)}\;\text{¬-r}
        }{\top \Rightarrow\; ^{\exists child}\neg Doctor, ^{\exists child}Lawyer, ^{\forall child}(Rich \sqcup Doctor)}\;\text{weak-r}
      }{\top \Rightarrow\; ^{\exists child}\neg Doctor, \exists child.Lawyer, ^{\forall child}(Rich \sqcup Doctor)}\;\text{∃-r}
    }{\top \Rightarrow \exists child.\neg Doctor, \exists child.Lawyer, ^{\forall child}(Rich \sqcup Doctor)}\;\text{∃-r}
  }{\top \Rightarrow (\exists child.\neg Doctor) \sqcup (\exists child.Lawyer), ^{\forall child}(Rich \sqcup Doctor)}\;\text{⊔-r}
}{\vdots}
$$

$$
\cfrac{
  \cfrac{
    \cfrac{
      \cfrac{
        \cfrac{
          \cfrac{
            \cfrac{
              \top \Rightarrow (\exists child.\neg Doctor) \sqcup (\exists child.Lawyer), ^{\forall child}(Rich \sqcup Doctor)
            }{^{\exists child}\top \Rightarrow\; ^{\exists child}((\exists child.\neg Doctor) \sqcup (\exists child.Lawyer)), ^{\exists child, \forall child}(Rich \sqcup Doctor)}\;\text{prom-}\exists
          }{^{\exists child}\top, ^{\forall child}\neg((\exists child.\neg Doctor) \sqcup (\exists child.Lawyer)) \Rightarrow\; ^{\exists child, \forall child}(Rich \sqcup Doctor)}\;\text{¬-l}
        }{^{\exists child}\top, ^{\forall child}\neg((\exists child.\neg Doctor) \sqcup (\exists child.Lawyer)) \Rightarrow\; ^{\exists child}\forall child.(Rich \sqcup Doctor)}\;\text{∀-r}
      }{^{\exists child}\top, \forall child.\neg((\exists child.\neg Doctor) \sqcup (\exists child.Lawyer)) \Rightarrow\; ^{\exists child}\forall child.(Rich \sqcup Doctor)}\;\text{∀-l}
    }{^{\exists child}\top, \forall child.\neg((\exists child.\neg Doctor) \sqcup (\exists child.Lawyer)) \Rightarrow \exists child.\forall child.(Rich \sqcup Doctor)}\;\text{∃-r}
  }{\exists child.\top, \forall child.\neg((\exists child.\neg Doctor) \sqcup (\exists child.Lawyer)) \Rightarrow \exists child.\forall child.(Rich \sqcup Doctor)}\;\text{∃-l}
}{\exists child.\top \sqcap \forall child.\neg((\exists child.\neg Doctor) \sqcup (\exists child.Lawyer)) \Rightarrow \exists child.\forall child.(Rich \sqcup Doctor)}\;\text{⊓-l}
$$

This proof tree could be explained by the following text:

(1) Doctors are Doctors or Rich (2) So, Everyone having all children Doctors has all children Doctors or Rich. (3) Hence, everyone either has at least a child that is not a doctor or every children is a doctor or rich. (4) Moreover, everyone is of the kind above, or, alternatively, have at least one child that is a lawyer. (5) In other words, if everyone has at least one child, then it has one child that has at least one child that is a lawyer, or at least one child that is not a doctor, or have all children doctors or rich. (6) Thus, whoever has all children not having at least one child not a doctor or at least one child lawyer has at least one child having every children doctors or rich.

The above explanation was built from top to bottom (toward the conclusion of the proof), by a procedure that tries not to repeat conjunctive particles (if—then, thus, hence, henceforth , moreover etc) to put together phrases derived from each subproof. In this case, phrase (1) comes from weak-r, ⊔-r; phrase (2) comes from prom-2; (3) is associated to weak-l, neg-r; (4) corresponds to weak-r, the two following ∃-r and the ⊓; (5) is associated to prom-1 and finally (6) corresponds to the remaining of the proof. The reader can note the large possibility of using endophoras in the construction of texts from structured proofs as the ones obtained by either SC$_{\mathcal{ALC}}$ or SC$^{[]}_{\mathcal{ALC}}$.

In Sect. 7.3 an example illustrating the use of theoremhood to explain reasoning on UML models is accomplished by proofs in **ND**, **SC** and **AT**.

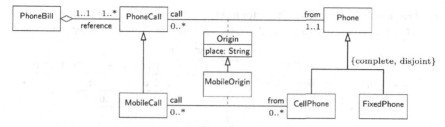

Fig. 7.4 UML class diagram

$$\text{Origin} \sqsubseteq \forall\text{place.String}$$

$$\text{Origin} \sqsubseteq \exists\text{place.}\top \sqcap (\leq 1 \text{ place})$$

$$\text{Origin} \sqsubseteq \exists\text{call.PhoneCall} \sqcap (\leq 1 \text{ call}) \sqcap \exists\text{from.Phone} \sqcap (\leq 1 \text{ from})$$

$$\text{MobileOrigin} \sqsubseteq \exists\text{call.MobileCall} \sqcap (\leq 1\text{call}) \sqcap \exists\text{from.CellPhone} \sqcap (\leq 1 \text{ from})$$

$$\text{PhoneCall} \sqsubseteq (\geq 1 \text{ call}^-.\text{Origin}) \sqcap (\leq 1 \text{ call}^-.\text{Origin})$$

$$\top \sqsubseteq \forall\text{reference}^-.\text{PhoneBill} \sqcap \forall\text{reference.PhoneCall}$$

$$\text{PhoneBill} \sqsubseteq (\geq 1 \text{ reference}^-)$$

$$\text{PhoneCall} \sqsubseteq (\geq 1 \text{ reference}) \sqcap (\leq 1 \text{ reference})$$

$$\text{MobileCall} \sqsubseteq \text{PhoneCall}$$

$$\text{MobileOrigin} \sqsubseteq \text{Origin}$$

$$\text{CellPhone} \sqsubseteq \text{Phone}$$

$$\text{FixedPhone} \sqsubseteq \text{Phone}$$

$$\text{CellPhone} \sqsubseteq \neg\text{FixedPhone}$$

$$\text{Phone} \sqsubseteq \text{CellPhone} \sqcup \text{FixedPhone}$$

Fig. 7.5 The $\mathcal{ALCQI}$ theory obtained from the UML diagram on Fig. 7.4

7.3 Explaining UML in ND$_{\mathcal{ALCQI}}$

In [2], DLs are used to formalize UML diagrams. It uses two DL languages: $\mathcal{DLR}_{ifd}$ or $\mathcal{ALCQI}$. The diagram on Fig. 7.4 and its formalization on Fig. 7.5, are from [2].

We use examples of DL deductions from [2, p. 84], using ND$_{\mathcal{ALCQI}}$ to reason on the $\mathcal{ALCQI}$ KB. The idea is to exemplify how one can obtain from ND$_{\mathcal{ALCQI}}$ proofs, a more precise and direct explanation.

The first example concerns a refinement of a multiplicity. That is, from reasoning on the diagram, one can deduce that the class MobileCall participates on the association MobileOrigin with multiplicity $0 \dots 1$, instead of the $0 \dots *$ presented in the diagram. The proof on ND$_{\mathcal{ALCQI}}$ is as follows, where we abbreviate the class names for their first letters, for instance, Origin (O), MobileCall (MC), call (c) and so on. Note that $\neg \geq 2c^-$.MO is actually an abbreviation for $\leq 1c^-$.MO.

$$
\cfrac{
  \cfrac{
    \cfrac{
      \text{MO} \sqsubseteq 0
      \qquad
      [\geq 2\,c^-.\text{MO}]^2
    }{
      \geq 2\,c^-.\text{MO} \sqsubseteq\, \geq 2\,c^-.0
    }
    \qquad
    \cfrac{
      \cfrac{[\text{MC}]^1 \quad \text{MC} \sqsubseteq \text{PC}}{\text{PC}}
      \qquad
      \text{PC} \sqsubseteq\, \geq 1\,c^-.0 \sqcap\, \leq 1\,c^-.0
    }{
      \cfrac{\geq 1\,c^-.0 \sqcap\, \leq 1\,c^-.0}{\leq 1\,c^-.0}
    }
  }{
    \cfrac{\bot}{\neg \geq 2\,c^-.\text{MO}}\;2
  }
}{
  \text{MC} \sqsubseteq \neg \geq 2\,c^-.\text{MO}
}\;1
$$

To exemplify deductions on diagrams, an incorrect generalization between two classes was introduced. The generalization asserts that each CellPhone is a FixedPhone, which means the introduction of the new axiom CellPhone $\sqsubseteq$ FixedPhone in the KB. From that improper generalization, several undesirable properties could be drawn.

The first conclusion about the modified diagram is that Cellphone is now inconsistent. The ND$_{\mathcal{ALCQI}}$ proof below explicits that from the newly introduced axiom and from the axiom CellPhone $\sqsubseteq$ ¬FixedPhone in the KB, one can conclude that CellPhone is now inconsistent.

$$
\cfrac{
  \cfrac{\text{Cell} \sqsubseteq \neg\text{Fixed} \quad [\text{Cell}]^1}{\neg\text{Fixed}}
  \qquad
  \cfrac{\text{Cell} \sqsubseteq \text{Fixed} \quad [\text{Cell}]^1}{\text{Fixed}}
}{
  \cfrac{\bot}{\text{Cell} \sqsubseteq \bot}\;1
}
$$

The second conclusion is in the modified diagram, Phone $\equiv$ FixedPhone. Note that we only have to show that Phone $\sqsubseteq$ FixedPhone since FixedPhone $\sqsubseteq$ Phone is an axiom already in the original KB. We can conclude from the proof below that Phone $\sqsubseteq$ FixedPhone is not a direct consequence of CellPhone being inconsistent, as stated in [2], but it is mainly as a direct consequence of the newly introduced axiom and a case analysis over the possible subtypes of Phone.

$$
\cfrac{
  \cfrac{[\text{Phone}]^1 \qquad \text{Phone} \sqsubseteq \text{Cell} \sqcup \text{Fixed}}{\text{Cell} \sqcup \text{Fixed}}
  \qquad
  \cfrac{[\text{Cell}] \qquad \text{Cell} \sqsubseteq \text{Fixed}}{\text{Fixed}}
  \qquad
  [\text{Fixed}]
}{
  \cfrac{\text{Fixed}}{\text{Phone} \sqsubseteq \text{Fixed}}\;1
}
$$

Below it is shown the above discussed subsumption proved in **SC** (Sequent Calculus).

$$
\cfrac{
  \cfrac{
    \cfrac{\text{MO} \Rightarrow 0}{\geq 2\,\text{call}^-.\text{MO} \Rightarrow\, \geq 2\,\text{call}^-.0}
  }{
    \text{MC}, \geq 2\,\text{call}^-.\text{MO} \Rightarrow\, \geq 2\,\text{call}^-.0
  }
  \qquad
  \cfrac{
    \text{MC} \Rightarrow \text{PC} \qquad \text{PC} \Rightarrow\, \geq 1\,\text{call}^-.0 \sqcap\, \leq 1\,\text{call}^-.0
  }{
    \cfrac{\text{MC} \Rightarrow\, \geq 1\,\text{call}^-.0 \sqcap\, \leq 1\,\text{call}^-.0}{\text{MC}, \geq 2\,\text{call}^-.\text{MO} \Rightarrow\, \geq 1\,\text{call}^-.0 \sqcap\, \leq 1\text{call}^-.0}
  }
}{
  \cfrac{
    \cfrac{\text{MC}, \geq 2\,\text{call}^-.\text{MO} \Rightarrow\, \geq 1\,\text{call}^-.0 \sqcap\, \leq 1\text{call}^-.0 \sqcap\, \geq 2\text{call}^-.0}{\text{MC}, \geq 2\,\text{call}^-.\text{MO} \Rightarrow \bot}
  }{
    \text{MC} \Rightarrow \neg \geq 2\,\text{call}^-.\text{MO}
  }
}
$$

In order to the reader concretely see that it is harder explaining on Tableaux basis than on Natural Deduction basis, we prove the same MC $\sqsubseteq \neg \geq 2$ call$^-$.MO subsumption in Tableaux. We follow [1, Section 2.3.2.1] and represent the Tableaux constraints as **ABox** assertions without unique name assumption.[2] The constraint"a belongs to (the interpretation of) C" is represented by $C(a)$ and "b is an R-filler of a" by $R(a, b)$. A complete presentation of the Tableaux procedure for $\mathcal{ALCQI}$ can be found at [1].

The Tableaux procedure starts translating the subsumption problem to a satisfiability problem. The subsumption $C \sqsubseteq D$ holds iff $C \sqcap \neg D$ is unsatisfiable. In our case, $C_0 \equiv \text{MC} \sqcap \geq 2\ \text{call}^-.\text{MO}$ should be unsatisfiable. Since C_0 is already in the NNF (negation normal form), we are ready to the Tableaux algorithm, otherwise we would have to first transform it to obtain a NNF equivalent concept description. Tableaux procedure starts with the ABox $A_0 = \{C_0(x_0)\}$ and applies consistency-preserving transformation rules to the ABox until no more rules apply. If the completed expanded ABox obtained does not contain clashes (contradictory assertions), then A_0 is consistent and thus C_0 is satisfiable, and incosistent (unsatisfiable) otherwise.

$\mathcal{A}_0$ is the initial ABox. By $\sqcap$-rule, we get $\mathcal{A}_1$. Than, by $\geq$-rule we get $\mathcal{A}_2$. $\mathcal{A}_3$ is obtained by using the theory axioms $\text{MO} \sqsubseteq \text{O}$ and $\text{MC} \sqsubseteq \text{PC}$. The ABox $\mathcal{A}_4$ is obtained by using the theory axiom $\text{PC} \sqsubseteq\ \geq 1\ \text{call}^-.\text{O} \sqcap\ \leq 1\ \text{call}^-.\text{O}$. Next, $\mathcal{A}_5$ by $\sqcap$-rule. ABox $\mathcal{A}_6$ now contains a contradiction, the individual a is required to have at most one successor of type O in the role call^-. Nevertheless, b and c are also required to be of type O and successors of a in role call^-, vide $\mathcal{A}_3$ and $\mathcal{A}_2$. This shows that C_0 is unsatisfiable, and thus $\text{MC} \sqsubseteq \neg \geq 2\ \text{call}^-.\text{MO}$.

$$\{(\text{MC} \sqcap \geq 2\ \text{call}^-.\text{MO})(a)\} \quad (\mathcal{A}_0)$$

$$\mathcal{A}_0 \cup \{\text{MC}(a), (\geq 2\ \text{call}^-.\text{MO})(a)\} \quad (\mathcal{A}_1)$$

$$\mathcal{A}_1 \cup \{\text{call}^-(a,b), \text{call}^-(a,c), \text{MO}(b), \text{MO}(c), a \neq b, b \neq c, a \neq c\} \quad (\mathcal{A}_2)$$

$$\mathcal{A}_2 \cup \{\text{O}(b), \text{O}(c), \text{PC}(a)\} \quad (\mathcal{A}_3)$$

$$\mathcal{A}_3 \cup \{(\geq 1\ \text{call}^-.\text{O} \sqcap\ \leq 1\ \text{call}^-.\text{O})(a)\} \quad (\mathcal{A}_4)$$

$$\mathcal{A}_4 \cup \{(\geq 1\ \text{call}^-.\text{O})(a), (\leq 1\ \text{call}^-.\text{O})(a)\} \quad (\mathcal{A}_5)$$

References

1. Baader, F.: The Description Logic Handbook: Theory, Implementation, and Applications. Cambridge University Press, Cambridge (2003)
2. Berardi, D., Calvanese, D., De Giacomo, G.: Reasoning on UML class diagrams. Artif. Intell. **168**(1–2), 70–118 (2005)
3. Maibaum, T.S.E.: The epistemology of validation and verification testing. In: Testing of Communicating Systems, 17th IFIP TC6/WG 6.1 International Conference, TestCom 2005, pp. 1–8. Montreal, Canada (2005)

---

[2] Instead, we allow explicit inequality assertions of the form $x \neq y$. Those assertions are assumed symmetric.

4. Prawitz, D.: Natural deduction: a proof-theoretical study. Ph.D. Thesis, Almqvist & Wiksell (1965)
5. Smullyan, R.: First-Order Logic. Springer, Berlin (1968)
6. Takeuti, G.: Proof Theory. Number 81 in Studies in Logic and the Foundations of Mathematics. North-Holland, Amsterdam (1975)

Chapter 8
A Prototype Theorem Prover

Reasoning is the ability to make inferences, and automated reasoning is concerned with the building of computing systems that automate this process.

Stanford Encyclopedia of Philosophy

Abstract In this chapter we present a prototype implementation of the systems $SC_{\mathcal{ALC}}$ and $SC_{\mathcal{ALCQ}}$. We choose to implement the Sequent Calculi because they represent a first step towards a ND implementations. The prototype theorem prover was implemented in Maude (Clavel et al. (2009) Maude manual (version 2.4). Technical Report, SRI International). So in Sect. 8.1 we present the Maude System and language and in Sect. 8.2 we describe the prototype implementation.

Keywords Deduction system · $\mathcal{ALC}$ · $\mathcal{ALCQI}$ · Maude · Proof search · Automatic deduction · Theorem prover

8.1 Overview of the Maude System

This section presents a general overview of the main characteristics of the Maude system and language. A complete description of Maude can be found at [3]. We will only present the aspects of Maude used in our implementation. Moreover, we will not present the theory foundations of Maude in the "Rewriting logic" [5] since our implementation uses the Maude system as an interpreter for the Maude language. We did not explore any possible mapping between description logics and rewriting logic.

Maude's basic programming statements are very simple and easy to understand. They are equations and rules, and have in both cases a simple rewriting semantics in

A. Rademaker, *A Proof Theory for Description Logics*,
SpringerBriefs in Computer Science, DOI: 10.1007/978-1-4471-4002-3_8,
© The Author(s) 2012

which instances of the lefthand side pattern are replaced by corresponding instances of the righthand side.

Maude programs are organized in modules. Maude modules containing only equations are called functional modules. Modules containing rules are called system module. In both cases, besides equations and rules, modules may contain declarations of sorts (types), operators and variables.

A functional module defines one or more functions by means of equations. Equations are used as simplification rules. Replacement of equals by equals is performed only from left to right as simplification rewriting. A function specification should have a final result and should be unique. Finally, Maude equations can be conditional, that is, they are only applied if a certain condition holds.

In a system module the rules are also computed by rewriting from left to right, but they are not equations. Instead, they are understood as local transition between states in a possibly concurrent system. For instance, a distributed banking system can be represented as account objects and messages floating in a "soup". That is, in a multi-set or bag of objects and messages. Such objects and messages in the soup can interact locally with each other according to specific rewrite rules. The systems specified by rules can be highly concurrent and nondeterministic. Unlike for equations, there is no assumption that all rewrite sequences will lead to the same outcome. Furthermore, for some systems there may not be any final states: their whole point may be to continuously engage in interactions with their environment as reactive systems. Note that, since the Maude interpreter is sequential, the concurrent behavior is simulated by corresponding interleavings of sequential rewriting steps. Logically, when rewriting logic was used as a logical framework to represent other logics a rule specifies a logical inference rule, and rewriting steps therefore represent inference steps.

Maude has two varieties of types: sorts, which correspond to well-defined data, and kinds, which may contain error elements. Sorts can be structured in subsort hierarchies, with the subsort relation understood semantically as subset inclusion. This allows support for partial functions, in the sense that a function whose application to some arguments has a kind but not a sort should be considered undefined for those arguments. Furthermore, operators can be subsort-overloaded, providing a useful form of subtype polymorphism.

In Maude the user can specify operators. An operator has arguments (each one has a sort) and a result sort. Each operator has its own syntax, which can be prefix, postfix, infix, or a "mixfix" combination. This is done by indicating with underscores the places where the arguments appear in the mixfix syntax. The combination of user-definable syntax with equations and equational attributes for matching leads to a very expressive capability for specifying any user-definable data. This is one of the main reasons that makes Maude a perfect language/system for prototyping.

Rewriting with both equations and rules takes place by matching a lefthand side term against the subject term to be rewritten. The most simpler matching is syntactic matching, in which the lefthand side term is matched as a tree on the (tree representation of the) subject term. Nevertheless, Maude also allows more expressive matching like "equational matching". when we define operators in Maude we can use attributes like assoc (associative) and comm (commutative) called equational attributes. For instance, if an operator is defined with both of these attributed, terms having this operator as the principal operator (the most external one), are not matching of trees, but as multi-set, that is, moduló associativity and commutativity. In general, a binary operator declared in a Maude can be defined with any combination of the equational attributes: associativity, commutativity, left-, right-, or two-sided identity, and idempotency.

A Maude system module implements a *rewrite theory* that must be *admissible*, which means that rules should be coherent relative to the equations [3]. If a rewrite theory contains both rules and equations, rewriting is performed modulo such equations. Maude strategy to rewriting terms is to first apply the equations to reach a canonical form, and then do a rewriting step with a rule (in a rule-fair manner). This strategy is complete if we assume coherence. Coherence means that we will not miss possible rewrites with rules that could have been performed if we had not insisted on first simplifying the term to its canonical form with the equations. Maude implicitly assumes this coherence property.

8.2 A Prototype Theorem Prover

In this section we present our Maude implementation of SC_{ALC} and $SC^{[]}_{ALC}$ sequent calculi. We will omit trivial details of the implementation and focus on the important parts. Moreover, it is important to note that this prototype is available for download at http://github.com/arademaker/SALC and also includes the implementation of SC_{ALCQI} system and its counterpart $SC^{[]}_{ALCQI}$. Those implementations are not described here since they do not differ considerably from the presented.

8.2.1 The Logical Language

Due to the flexibility to specify user-definable data in Maude, the definition of the description logics ALC and $ALCQI$ syntax was effortless.

The language ALC is defined in the function module SYNTAX below. We have defined sorts for atomic concepts and atomic roles besides the sort for concepts and roles in general. The constants $\top$ and $\bot$ were also specified.

```
fmod SYNTAX is
 inc NAT .

 sorts AConcept Concept ARole Role .
 subsort AConcept < Concept .
 subsort ARole < Role .
 ops ALL EXIST : Role Concept -> Concept .
 ops CTRUE CFALSE : -> AConcept .
 op _&_ : Concept Concept -> Concept [ctor gather (e E) prec 31] .
 op _|_ : Concept Concept -> Concept [ctor gather (e E) prec 32] .
```

```
op ~_ : Concept -> Concept [ctor prec 30] .

  eq ~ CTRUE = CFALSE .
  eq ~ CFALSE = CTRUE .
endfm
```

The syntax for defining operators is:

```
op NAME : Sort-1 Sort-2 ... -> Sort [attr-1 ...] .
```

where NAME may contain underscores to identify arguments position in infix notation. The list of sorts before -> is the arguments and the sort after is the sort of the resultant term.

Since our SC_{ALC} and SC_{ALCQI} systems reason over labeled concepts. The next step was to extend the language with labels and some functions over them. A labeled concept $^{\forall R, \exists S}\alpha$ is represented by the term < al(R) ex(S) | A > where A is a constant of the sort AConcept and R and S constants of the sort ARole. In the modules below, we show the declarations of all operators but omitted the specification of logical operators has-quant, has-lt and so on.

```
fmod LABEL is
  inc SYNTAX .

  sorts Label ELabel ALabel QLabel .
  subsorts ELabel ALabel QLabel < Label .

  ops gt lt : Nat Role -> QLabel .
  op ex : Role -> ELabel .
  op al : Role -> ALabel .
endfm
```

The definition below of the operators neg and neg-aux should be clear but being the first equational specification deserves an explanation. The operator neg(L) operates over the list of labels L inverting all its quantifiers. In Sect. 3.1, we represent such operation as $\neg L$. We use neg-aux to interact over the list accumulating the result in its second argument until the first argument is completely consumed and the second argument returned.

```
fmod LALC-SYNTAX is
  inc LABEL .
  inc LIST{Label} .

  vars L1 L2 : List{Label} .
  vars R : Role .
  var C : Concept .
```

```
sorts Expression LConcept .
subsort LConcept < Expression .

op <_|_> : List{Label} Concept -> LConcept [ctor] .

ops has-quant has-lt has-gt : List{Label} -> Bool .
ops has-al has-ex : List{Label} -> Bool .
op neg : List{Label} -> List{Label} .
op neg-aux : List{Label} List{Label} -> List{Label} .
...
eq neg(L1) = neg-aux(L1, nil) .

eq neg-aux(L1 al(R), L2) = neg-aux(L1, ex(R) L2) .
eq neg-aux(L1 ex(R), L2) = neg-aux(L1, al(R) L2) .
eq neg-aux(nil, L2) = L2 .
endfm
```

It is worth noting that this is not the only way to define neg in Maude, the auxiliary function is not necessary at all, but we will use them frequently in our implementation.

Finally, the module LALC-SYNTAX declares the sorts Expression and LConcept (labeled concept). Expressions are labeled concepts but the distinction can be useful for future extensions of the calculi.

8.2.2 The Sequent Calculus

In the function module SEQUENT-CALCULUS we implemented the generic data structures that are used by all sequent calculi. The idea is that a proof will be represented as a multi-set ("soup") of goals and messages (operators with sort State). Goals are sequents with additional properties to keep the proof structure. Each goal will have an identifier (natural number), the goal origin, the name of the rule used to produce that goal, and the sequent. In this way, our proof is a graph represented as a multi-set of terms with sort Proof. The goals operator holds a list of natural numbers as its argument, the list of pending goals. The next operator is just an auxiliary operator that provides in each proof step the next goal identifier.

```
fmod SEQUENT-CALCULUS is
 inc LALC-SYNTAX .
 inc SET{Expression} .
 inc SET{Label} .
 ...
 sorts Sequent Goal State Proof .
 subsort Goal State < Proof .

 op next  : Nat -> State .
```

```
op goals : Set{Nat} -> State .

op [_from_by_is_] : Nat Nat Qid Sequent -> Goal [ctor] .

op nil : -> Proof [ctor] .
op __ : Proof Proof -> Proof [ctor comm assoc] .

op _|-_ : Set{Expression} Set{Expression} ->
          Sequent [ctor prec 122 gather(e e)] .

op _:_|-_:_ : Set{Expression} Set{Expression} Set{Expression}
              Set{Expression} ->
              Sequent [ctor prec 122 gather (e e e e)] .
...
endfm
```

We must also note that we have defined two operators[1] to construct sequents. The operator $\_|-\_$ is the simplest sequent with two multi-set of expression, one on the left (sequent antecedent, possibly empty) and other on the right (sequent succedent, possibly empty), it is used to implement $SC_{\mathcal{ALC}}$. The operator $\_:\_|-\_:\_$ is used by the frozen versions of $SC_{\mathcal{ALC}}$ and $SC_{\mathcal{ALCQI}}$. The two additional external sets of expressions hold the frozen formulas.

Consider the proof of the sequent $\forall R.(A \sqcap B) \Rightarrow \forall R.A \sqcap \forall R.B$ presented in Fig. 8.1. One proof constructed by our system is represented by the term below. The goal 0 is the initial state of the proof, goals 6 and 5 are the initial sequents. Goal 1 is obtained from goal 0 applying the rule $\forall$-l. The empty argument of goals (empty) represents the fact that this proof is complete, there is no remaining goals to be proved.

```
goals(empty) next(7)
[0 from 0 by 'init is < nil | ALL(R, A & B) > |-
                       < nil | ALL(R, A) & ALL(R, B) >]

[1 from 0 by 'forall-l is < al(R) | A & B > |-
                           < nil | ALL(R, A) & ALL(R, B) >]

[2 from 1 by 'and-l is < al(R) | A >, < al(R) | B > |-
                       < nil | ALL(R, A) & ALL(R, B) >]

[3 from 2 by 'and-r is < al(R) | A >, < al(R) | B > |- < nil | ALL(R, A) >]
[4 from 2 by 'and-r is < al(R) | A >, < al(R) | B > |- < nil | ALL(R, B) >]
[5 from 3 by 'forall-r is < al(R) | A >, < al(R) | B > |- < al(R) | A >]
[6 from 4 by 'forall-r is < al(R) | A >, < al(R) | B > |- < al(R) | B >]
```

Fig. 8.1 An example of a proof in the implementation of $SC_{\mathcal{ALC}}$

---

[1] Term constructor in Maude terminology since these operators will never be reduced, they are used to hold data.

8.3 The SC$_{\mathcal{ALC}}$ System

The SC$_{\mathcal{ALC}}$ system was implemented in a system module. Basically, each rule of
the system is a Maude rewriting rule. The rewriting procedure construct the proof
bottom-up.

```
mod SYSTEM is
 inc SEQUENT-CALCULUS .

 [rules and equations presented below]

endm
```

The first observation regards the structural rules of SC$_{\mathcal{ALC}}$. Since the left and
right sides of the sequents are sets of formulas, we do not need permutation of
contraction rules. We also proved in Sect. 3.4 that the cut rule was not necessary
either. Nevertheless, we could lose completeness if we have omitted the weak rules.
We need them to allow the promotional rules applications. Moreover, the initial
sequent was implemented as an equation rather than as a rule. We used the fact that
in Maude all rewriting steps with rules are executed module equational reductions.
The implementation of the initial sequents using equations means that a goal detected
as initial will be removed from the goals lists right away.

```
eq [ X from Y by Q is ALFA, E |- E, GAMMA ] goals((X, XS)) =
   [ X from Y by Q is ALFA, E |- E, GAMMA ] goals((XS))
 [label initial] .

rl [weak-l] :
 [ X from Y by Q is ALFA, E |- GAMMA ] next(N) goals((X, XS))
 =>
 [ X from Y by Q is ALFA, E |- GAMMA ] next(N + 1)
 goals((XS, N))
 [ N from X by 'weak-l is ALFA |- GAMMA ] .
```

First we note the difference between rules and equations. They are very similar
expected that the former uses => and the later = as a term separator.

```
rl [label] : term-1 => term-2 [attr-1,...] .
eq term-1 = term-2 [attr-1,...] .
```

We note that on each rule the goal being rewritten must be repeated in the left and
right side of the rule. See weak rule above. If we omit the goal on the right side of
the rule we will be removing the goal from the proof. We are actually including new
goals on each step, that is, we put new goals in the "soup" of goals.

Reading bottom-up, some rules create more than one (sub)-goal from a goal. This
is the case of rule ⊓-r below. Besides that, whenever a rule has some additional

proviso, we use Maude *conditional rules* to express the rule proviso in the rule condition. In the rule ⊓-r, the proviso states that in the list of labels of the principal formula all labels must be universal quantified, in $SC_{\mathcal{ALC}}$, this is the same of saying that L cannot contain existential quantified labels (has-ex(L)).

```
crl [and-r] :
  [ X from Y by Q is ALFA |- GAMMA, < L | A & B > ]
  next(N) goals((X, XS))
  =>
  next(N + 2) goals((XS, N, N + 1))
  [ X from Y by Q is ALFA |- GAMMA, < L | A & B > ]
  [ N     from X by 'and-r is ALFA |- GAMMA, < L | A > ]
  [ N + 1 from X by 'and-r is ALFA |- GAMMA, < L | B > ]
if not has-ex(L) .
```

The rule condition can consist of a single statement or can be a conjunction formed with the associative connective /\. Rule promotional-∃ has two conditions. The first, from left to right, is the rule proviso (all concepts on the left-side of the sequent must have the same most external label), the second is actually just an instantiation of the variable GAMMA' with the auxiliary operator remove-label. GAMMA' will be the right-side of the new sequent (goal) created. remove-label iterate over the concepts removing the most external label of them.

```
crl [prom-exist] :
  [ X from Y by Q is < ex(R) L | A > |- GAMMA ]
  next(N) goals((X, XS))
  =>
  next(N + 1) goals((XS, N))
  [ X from Y by Q is < ex(R) L | A > |- GAMMA ]
  [ N from X by 'prom-exist is  < L | A > |- GAMMA' ]
if all-label(GAMMA, ex(R)) = true
   /\ GAMMA' := remove-label(GAMMA, ex(R), empty) .
```

The implementation of the remain rules is straightforward. We have one observation more about the rules above, the argument of next(N) gives the next goal identifier. The argument of goals holds the list of goals not solved. A derivation with goals(empty) in the "soup" is a completed proof of the sequent in the goal with identifier 0.

8.3.1 The $SC^{\square}_{\mathcal{ALC}}$ System Implementation

The system $SC^{\square}_{\mathcal{ALC}}$ is implemented in a very similar way of $SC_{\mathcal{ALC}}$. The main differences are that sequents now have frozen concepts and two additional rules had to be implemented. Concepts that were frozen together will never be unfrozen separated, so that, instead of defining an operator to freeze a concept, we defined a constructor of a set of frozen concepts.

```
mod SYSTEM is
inc SEQUENT-CALCULUS .
...
op [_,_,_] : Nat Nat Set{Expression} -> Expression .
```

The constructor of frozen set of concepts has three arguments. The first argument is the context identifier (see Sect. 4.3) created to group the pair of sets of concepts frozen together on the sequent antecedent and succedent. The second argument is the state of the context where 0 means that the context is saved but not reduced yet (context was frozen by weak rule), and 1 means that the context was reduced (context was frozen by frozen-exchange rule). The last argument is the set of frozen concepts.

Almost all rules of SC$^{[]}_{\mathcal{ALC}}$ do not touch in the frozen concepts. This is the case of negation rule below. We note the use of the operator neg inverting the list of labels of a concept.

```
rl [neg-l] :
  [ X from Y by Q is
       FALFA : ALFA, < L | ~ A > |- GAMMA : FGAMMA ]
  next(N) goals((X, XS))
  =>
  next(N + 1) goals((XS, N))
  [ X from Y by Q is
       FALFA : ALFA, < L | ~ A > |- GAMMA : FGAMMA ]
  [ N from X by 'neg-l is
       FALFA : ALFA |- GAMMA, < neg(L) | A > : FGAMMA ] .
```

The weak-r rule was implemented as a conditional rewrite rule below. The left and right-side of the sequent in goal X were frozen and added to the set of frozen concepts on the left and right side of the sequent in the new goal N. The variables FALFA and FGAMMA match the set of frozen concepts on both sides. The weak-l rule is similar.

```
crl [weak-r] :
  [ X from Y by Q is FALFA : ALFA |- GAMMA, E : FGAMMA ]
  next(N) goals((X, XS))
  =>
  next(N + 1) goals((XS, N))
  [ X from Y by Q is FALFA : ALFA |- GAMMA, E : FGAMMA ]
  [ N from X by 'weak-l is
     (FALFA, [M:Nat, 0, ALFA]) : ALFA |-
     GAMMA : (FGAMMA, [M:Nat, 0, (GAMMA, E)]) ]
  if M:Nat  := next-frozen(union(FALFA, FGAMMA)) .
```

The other SC$^{[]}_{\mathcal{ALC}}$ rule that modifies the set of frozen concepts in a goal is the frozen-exchange rule. The Maude pattern matching mechanism was very useful in the implementation of this rule. The rule select randomly[2] a context (sets of frozen concepts) to unfreeze – [O:Nat, 0, ES1] and [O:Nat, 0, ES2] – and freeze the set of formulas that are in the current context – ALFA and GAMMA. The pattern also guarantee that only contexts saved but not already reduced (second argument equals zero) will be selected. The new context created in the goal N has the second argument equals one – it is a reduced context. Maude's pattern matching mechanism is very flexible and powerful. On the other hand, this rule does not provide much control over the choice of contexts (set of frozen formulas) that will be unfrozen. This choice can have huge impact in the performance of a proof construction.

```
crl [frozen-exchange] :
  [ X from Y by Q is
        [O:Nat,0,ES1], FALFA : ALFA |-
        GAMMA : FGAMMA, [O:Nat,0,ES2] ]
  goals((X, XS)) next(N)
  =>
  goals((XS, N)) next(N + 1)
  [ X from Y by Q is
        [O:Nat,0,ES1], FALFA : ALFA |-
        GAMMA : FGAMMA, [O:Nat,0,ES2] ]
  [ N from X by 'frozen-exchange is
     ([M:Nat,1,ALFA], FALFA) : ES1 |-
     ES2 : (FGAMMA, [M:Nat,1,GAMMA]) ]

  if M:Nat := next-frozen(union(([O:Nat,0,ES1], FALFA),
                                ([O:Nat,0,ES2], FGAMMA))) .
```

8.3.2 The Interface

The current user interface of the prototype is the Maude prompt. We do not provide any high level user interface yet, although different alternatives exist for it. For example, we could implement the DIG [1] interface using Maude external objects [3]. The system module THEOREM-PROVER is the main interface with the prototype. It basically declares some constants of the sort AConcept (atomic concepts) and ARole (atomic roles) and the operator th_end. This operator is a "syntax sugar" to assist the user in the creation of the proof term in its initial state ready to be rewritten.

```
mod THEOREM-PROVER is
inc SYSTEM .

ops A B C D E : -> AConcept .
ops R S T U V : -> ARole .
```

---

[2] The selection is made by pattern matching of a context module commutative and associative, thanks to the attributes of the operator comma, the constructor of Set{Expression} terms.

```
op th_end : Sequent -> Goal .

vars ALFA GAMMA : Set{Expression} .
var SEQ : Sequent .

eq th SEQ end =
   [ 0 from 0 by 'init is SEQ ] next(1) goals(0) .
endm
```

The module THEOREM-PROVER includes the module SYSTEM, where SYSTEM can be any of the implemented systems presented in the previous sections.

With the help of the above module we can prove the theorem from Example 1 (8.1) using two alternatives.

$$\exists child.\top \sqcap \forall child.\neg(\exists child.\neg Doctor) \sqsubseteq \exists child.\forall child.Doctor \qquad (8.1)$$

We can use the already declared constants assuming $A = Doctor$ and the role $R = child$ or we can declare two new constants in a module that imports THEOREM-PROVER.

```
mod MY-TP is
  inc THEOREM-PROVER .

  op child : -> ARole .
  op Doctor : -> AConcept .
endm
```

In the second case, after entering the module MY-TP in Maude, we could test the proof initialization with the Maude command reduce (red). The command rewrite the given term using only equations. In that case, only the equation of the operator th_end from module THEOREM-PROVER is applied.

```
Maude> red
  th < nil | EXIST(child, CTRUE) &
             ALL(child, ~ EXIST(child, ~ Doctor)) > |-
   < nil | EXIST(child, ALL(child, Doctor)) > end .

result Proof: next(1) goals(0)
[0 from 0 by 'init is
   < nil |
     EXIST(child, CTRUE) &
     ALL(child, ~ EXIST(child, ~ Doctor)) >
   |-
   < nil | EXIST(child, ALL(child, Doctor)) > ]
```

To construct a proof of a given sequent, we can use Maude rewrite or search command. The former will return one possible sequence of rewriting steps until a *canonical term*[3] is reached. The latter will search for all possible paths of rewriting steps from the given initial state until the final given state.

Below we present the same sequent with *Doctor* and *child* replaced by *A* and *R* respectively. As we can see, due the presence of weak rules and the lack of a strategy to control the applications of the rules, we failed to obtain a proof for a valid sequent using the command `rewrite`.

```
Maude> rew th
    < nil | EXIST(R, CTRUE) & ALL(R, ~ EXIST (R, ~ A)) > |-
    < nil | EXIST(R, ALL(R, A)) > end .

result Proof: next(3) goals(2)
[0 from 0 by 'init is
    < nil | EXIST(R, CTRUE) & ALL(R, ~ EXIST (R, ~ A)) > |-
    < nil | EXIST(R, ALL(R, A)) >]
 [1 from 0 by 'weak-1 is empty |-
   < nil | EXIST(R, ALL(R, A)) >]
 [2 from 1 by 'weak-r is empty |- empty]
```

The `rewrite` command explores just one possible sequence of rewrites of a system described by a set of rewrite rules and an initial state. The search command allows one to explore (following a breadth-first strategy) the reachable state space in different ways.

Using the `search` command we can ask for all possible proof trees that can be constructed for a given sequent. Moreover, we can limit the space search with the two optional parameters [n,m] where *n* providing a bound on the number of desired solutions and *m* stating the maximum depth of the search. The search arrow `=>!` indicates that only canonical final states are allowed, that is, states that cannot be further rewritten. On the left-hand side of the search arrow we have the starting term, on the right-hand side the pattern that has to be reached, in the case below, `P:Proof goals(empty)`.

```
Maude> search [1,20]
        th < nil | EXIST(R, CTRUE) & ALL(R, ~ EXIST (R, ~ A)) >
                |- < nil | EXIST(R, ALL(R, A)) > end
        =>! P:Proof goals(empty) .

P:Proof --> next(10)
[0 from 0 by 'init is
 < nil | EXIST(R, CTRUE) & ALL(R, ~ EXIST (R, ~ A)) > |-
 < nil | EXIST(R, ALL(R, A)) >]
 [1 from 0 by 'and-1 is < nil | ALL(R, ~ EXIST (R, ~ A)) >,
 < nil | EXIST(R, CTRUE) > |-
 < nil | EXIST(R, ALL(R, A)) >]
  [2 from 1 by 'forall-1 is < nil | EXIST(R, CTRUE) >,
  < al(R) | ~ EXIST(R, ~ A) > |-
  < nil | EXIST(R, ALL(R, A)) >]
   [3 from 2 by 'neg-1 is < nil | EXIST(R, CTRUE) > |-
```

---

[3] A term that cannot be further rewritten.

```
< nil | EXIST(R, ALL(R, A)) >, < ex(R) | EXIST(R, ~ A) >]
[4 from 3 by 'exist-r is < nil | EXIST(R, CTRUE) > |-
< ex(R) | ALL(R, A) >, < ex(R) | EXIST(R, ~ A) >]
[5 from 4 by 'forall-r is < nil | EXIST(R, CTRUE) > |-
 < ex(R) | EXIST(R, ~ A) >, < ex(R) al(R) | A >]
[6 from 5 by 'exist-r is < nil | EXIST(R, CTRUE) > |-
 < ex(R) ex(R) | ~ A >, < ex(R) al(R) | A >]
[7 from 6 by 'exist-l is
 < ex(R) | CTRUE > |- < ex(R) ex(R) | ~ A >,
 < ex(R) al(R) | A >]
[8 from 7 by 'prom-exist is
 < nil | CTRUE > |-
 < ex(R) | ~ A >, < al(R) | A >]
[9 from 8 by 'neg-r is
 < nil | CTRUE >, < al(R) | A > |- < al(R) | A >]
```

Above, the variable P in the input pattern was bound in the result to the desired proof term, that is, the one with `goals(empty)`. Since P was the only variable in the pattern, the result shows only one binding. In other worlds, search results are bindings for variables in the pattern given after the search arrow.

Distributed with our prototype there is a simple Maude-2-LATEX proof terms translator developed by Caio Mello.[4] The translator receives as input a term like the one above and return its representation in LATEX using the LATEX package `bussproof` [2]. The output in LATEX is:

$$\cfrac{\cfrac{\cfrac{\cfrac{\cfrac{\cfrac{\cfrac{\top, {}^{\forall R}A \Rightarrow {}^{\forall R}A}{\top \Rightarrow {}^{\exists R}\neg A, {}^{\forall R}A} \scriptstyle{\neg\text{-r}}}{{}^{\exists R}\top \Rightarrow {}^{\exists R, \exists R}\neg A, {}^{\exists R, \forall R}A} \scriptstyle{\text{prom-}\exists}}{\exists R.\top \Rightarrow {}^{\exists R, \exists R}\neg A, {}^{\exists R, \forall R}A} \scriptstyle{\exists\text{-l}}}{\exists R.\top \Rightarrow {}^{\exists R}\exists R.(\neg A), {}^{\exists R, \forall R}A} \scriptstyle{\exists\text{-r}}}{\exists R.\top \Rightarrow {}^{\exists R}\exists R.(\neg A), {}^{\exists R}\forall R.A} \scriptstyle{\forall\text{-r}}}{\exists R.\top \Rightarrow {}^{\exists R}\exists R.(\neg A), \exists R.\forall R.A} \scriptstyle{\exists\text{-r}}}{\cfrac{\cfrac{\exists R.\top, {}^{\forall R}\neg \exists R.(\neg A) \Rightarrow \exists R.\forall R.A}{\exists R.\top, \forall R.(\neg \exists R.(\neg A)) \Rightarrow \exists R.\forall R.A} \scriptstyle{\forall\text{-l}}}{\exists R.\top \sqcap \forall R.(\neg \exists R.(\neg A)) \Rightarrow \exists R.\forall R.A} \scriptstyle{\sqcap\text{-l}}}$$

8.3.3 Defining Proof Strategies

An automated theorem prover would not be efficient or even useful if we cannot provide strategies for deduction rules applications. Moreover, from Sect. 4.3 we know that SC$^{[]}{}_{\mathcal{ALC}}$ deduction rules were designed to be used in a very specific strategy. Maude support two ways to define strategies for rewriting rules application. The first

---

[4] An undergraduate student working at TecMF/PUC-Rio Lab.

option is the original one, we can use Maude reflection feature to control of rules applications at the metalevel developing a full user-definable internal strategies. The second options is to use the Maude Strategy Language [4].

The strategy language allows the definition of strategy expressions that control the way a term is rewritten. The strategy language was designed to be used at the object level, rather than at the metalevel. There exist a strict separation between the rewrite rules in system modules and the strategy expressions, that are specified in separate strategy modules. Moreover, a strategy is described as an operation that, when applied to a given term, produces a set of terms as a result, given that the process is nondeterministic in general. In the current version of Maude, not all features of the strategy language are available in *Core* Maude. To be more precise, the *Core* Maude does not support recursive strategies. Recursion is achieved by giving a name to a strategy expression and using this name in the strategy expression itself or in other related strategies. Given that limitation, we use the prototype strategy language implementation in *Full* Maude [3].

In our current prototype version we defined the strategy described in Sect. 4.3 to control $SC^{[]}_{\mathcal{ALC}}$ rules applications. The basic strategies consist of the application of a rule (identified by the corresponding rule label) to a given term. Strategies operators allow the construction of complex strategy expressions.

The strategy expand presented below controls how the rules of $SC^{[]}_{\mathcal{ALC}}$ ought to be applied. It can be interpreted as: the system must first try to reduce the given term using one of the promotional rules (the union operator is |). If it is successful, the system must try to further transform the result term using $\forall$-{l,r}, $\exists$-{l,r}, $\sqcup$-{l,r}, $\sqcap$-{l,r} or $\neg$-{l,r} (the operator ; the a concatenation). If neither the promotional rules nor the previous mentioned rules could be applied, one of the weak rules should be tried. If none of the previous rules could be applied, the frozen-exchange rule must be tried.

```
(smod BACKTRACKING-STRAT is

strat solve : @ Proof .
strat expand : @ Proof .
var P : Proof .

sd expand := (((try(prom-exist | prom-all) ;
               (and-l | and-r | or-l | or-r | forall-l |
                forall-r | exist-l | exist-r |
                neg-l | neg-r))
              orelse (weak-l | weak-r))
             orelse frozen-exchange) .

sd solve := if (match P s.t. (is-solution(P))) then
       idle
    else
        expand ; if (match P s.t. (is-ok(P))) then
          solve else idle fi
     fi .
endsm)
```

The strategy expand defines how each proof step will be performed. The solve strategy is the complete strategy to construct a proof. It is basically a backtracking procedure, on each step, the system verifies if it has already a solution—using the defined operator is-solution. If the term is not a solution, it executes the expand step and check if the result term is a valid term, that is, a term still useful to reach to a solution—this is done with the operator is-ok. If the term is still valid but not yet a solution it continues recursively.

The implementations of is-solution and is-ok were done in a separated module. The operator is-ok evaluates to false whenever we detected a loop in the proof construction. There are differents loop situations, below we present one of them, when we have a sequent with two equal sets of frozen formulas (contexts).

```
op is-ok : Proof -> Bool .
op is-solution : Proof -> Bool .

eq is-solution(P:Proof goals(empty)) = true .
eq is-solution(P:Proof) = false [owise] .
...
eq is-ok(P:Proof
        [M from N by RL is FALFA1, [X1, X3, FALFA0],
            [X2, X4, FALFA0] : ALFA |- GAMMA :
                [X1, X3, FGAMMA0], [X2, X4, FGAMMA0],
                FGAMMA1])
    = false .

eq is-ok(P:Proof) = true [owise] .
```

Using the solve strategy defined above, we can prove the subsumption from Equation (8.1) in SC$^{[]}_{\mathcal{ALC}}$. We use the strategy aware command srew instead of the rew. In additional, since we are not using Full Maude, the command in Maude prompt is inside parentheses.

```
Maude> (srew th empty : < nil | EXIST(R, CTRUE) &
                            ALL(R, ~ EXIST (R, ~ A)) > |-
            < nil | EXIST(R, ALL(R, A)) > : empty end
        using solve .)

result Proof :
 goals(empty)next(10)
 [0 from 0 by 'init is
    empty : < nil | EXIST(R,CTRUE) & ALL(R,~ EXIST (R,~ A))>
   |- < nil | EXIST(R,ALL(R,A))> : empty]
 [1 from 0 by 'and-1 is
    empty : < nil | ALL(R,~ EXIST(R,~ A))>,
    < nil | EXIST(R,CTRUE)> |-
    < nil | EXIST(R,ALL(R,A))> : empty]
 [2 from 1 by 'forall-1 is
    empty : < nil | EXIST(R,CTRUE)>,
    < al(R) | ~ EXIST(R,~ A)> |-
```

```
              < nil | EXIST(R,ALL(R,A))> : empty]
     [3 from 2 by 'exist-l is empty : < al(R) | ~ EXIST (R,~ A)>,
         < ex(R) | CTRUE > |- < nil | EXIST(R,ALL(R,A))> : empty]
     [4 from 3 by 'exist-r is empty : < al(R) | ~ EXIST (R,~ A)>,
         < ex(R) | CTRUE > |- < ex(R) | ALL(R,A)> : empty]
     [5 from 4 by 'forall-r is empty : < al(R) | ~ EXIST (R,~ A)>,
         < ex(R) | CTRUE > |- < ex(R)al(R) | A > : empty]
     [6 from 5 by 'neg-l is
        empty : < ex(R) | CTRUE > |-
        < ex(R) | EXIST(R,~ A)>, < ex(R)al(R) | A > : empty]
     [7 from 6 by 'prom-exist is
        empty : < nil | CTRUE > |-
        < nil | EXIST(R,~ A)>, < al(R) | A > : empty]
     [8 from 7 by 'exist-r is
        empty : < nil | CTRUE > |- < al(R) | A >,
        < ex(R) | ~ A > : empty]
     [9 from 8 by 'neg-r is
        empty : < nil | CTRUE >, < al(R) | A >
        |- < al(R) | A > : empty]
```

References

1. Bechhofer, S., Patel-Schneider, P.F.: DIG 2.0: The DIG description logic interface overview (2006)
2. Buss, S.: The "buss proofs" latex/tex style file for creating proof trees (2006). http://math.ucsd.edu/ sbuss/ResearchWeb/bussproofs/index.html
3. Clavel, M., Durán, F., Eker, S., Lincoln, P., Martí-Oliet, N., Meseguer, J., Talcott, C.: Maude manual (version 2.4). Techical Report, SRI, International (2009)
4. Eker, S., Martí-Oliet, N., Meseguer, J., Verdejo, A.: Deduction, strategies, and rewriting. Electron. Notes Theoret. Comput. Sci. **174**(11), 3–25 (2007)
5. Meseguer, J.: Conditional rewriting logic as a unified model of concurrency. Theoret. Comput. Sci. **96**(1), 73–155 (1992)

Chapter 9
Conclusion

Abstract Description Logics have well-known and mature proof procedures based on Tableaux for reasoning on Ontologies and Knowledge Bases. The task of understanding the outcomes of formal proof procedure or consistency tests is sometimes quite hard. Explanations on the reasons for some subsumptions either hold or not are demanding. The latter is in general supported by a human-readable translation of the witness construction obtained by the usual, first-order inspired, Tableaux DL procedure. For the former, however, an explanation should be obtained from the proof resulted by this very Tableaux procedure. In this chapter, we review our contributions and present some possible future works.

Keywords Conclusion · Contribution · $\mathcal{ALC}$ · $\mathcal{ALCQI}$ · Justification · Proof theory · Proof explanation

9.1 Contributions

Considering the logical motivation of providing a purely propositional (not based on nominals) proof procedure for propositional DLs, we show two Sequent Calculus and two Natural Deductions defined by purely propositional terms. Considering the concrete use of DL reasoners, we believe that the use of a system that allows the use of non-analytic cuts (non-atomic cuts) is interesting whenever one takes into account the super-polynomial size of some cut-free proofs (such as the Pigeonhole Principle). Besides that, producing proofs of subsumptions inside a **TBOX**, without making use of the terminological gap imposed by the traditional Tableaux procedure, seems to an interesting step towards better explanation generations.

The main contributions of our work are twofold. Firstly, from the point of view of producing short proofs, we define proof systems that are able to produce proofs or derivations with cuts ($SC_{\mathcal{ALC}}$, $SC^{[]}_{\mathcal{ALC}}$ and its extension for $\mathcal{ALCQI}$) as well as non-normal proofs ($ND_{\mathcal{ALC}}$ and its extension for $\mathcal{ALCQI}$). The elimination of the cut rule as well as the normalization theorem are mandatory proof-obligations performed in this work aiming to prove that the systems are minimally mechaniz-able. The other contribution made in this work relies on the fact that the Sequent Calculus as well as the Natural Deduction are not strongly based on first-order mech-anisms and interpretations as the known Tableaux procedures are. The systems are purely propositional. In order to achieve this feature, a strong use of labeled formu-las is made. Thus, both, the Sequent Calculus and the Natural Deduction are labeled deductive systems, following the tradition initiated by Dov Gabbay [6]. Both features are steps towards the possibility of generating quite human-readable explanations. Besides those previous mentioned contribuitions we think that presenting an alter-native proof procedure for a well-know logic is a contribution in its own.

It is worth mentioning that the deduction systems presented in this book and, more generally, the proof-theoretical study on DLs, have directly aplications in the so called "justification", term introduced by the Semantic Web community. A justification is the minimum subset of an ontology (a set of non-logical axioms, TBox) sufficient for an entailment, regardless of entailment type being a subsumption of concepts or concept satisfiability test, to hold. Justification is currently the prevalent form of explanation in OWL ontology development tools. Of course, debugging and repair of an ontology is a crucial step in the ontology development process in order to ensure the quality and the correctness. Currently research on justification deal with the improvement of the comprehension of a single justification for an individual entailment [10, 11], performance improvement of computing justifications [2, 12] and handle of multiple justifications of a single entailment [3].

From a proof-theoretical perspective, the justification is the non logical top most axioms used in a proof of a entailment. This set is easily obtained in a bottom-up proof construction using the one of the sequent calculi presented in this book.

Regarding the Natural Deduction systems presented for $\mathcal{ALC}$ and $\mathcal{ALCQI}$, despite providing a variation of themes, the main motivation is the possibility of getting ride on a weak form of the Curry-Howard isomorphism in order to provide explanations with greater content. This last affirmative takes into account that the reading (explanatory) content of a proof is a direct consequence of its computational content.

We not only presented ND systems for $\mathcal{ALC}$ and $\mathcal{ALCQI}$ but also showed, by means of some examples, how they can be useful to explain formal facts on theories obtained from UML models. Instead of UML, ER could also be used according a similar framework. Regarding the examples used and the explanations obtained, it is worthwhile noting that the Natural Deduction proofs obtained are quite close to the natural language explanation provided. It is a future task to provide the respective natural language explanation for a comparison. We aimed to show that **ND** deduction systems are better than Tableaux and Sequent Calculus as structures to be used in explaining theorem when validating theories in the presence of false positives. That is, when a valid subsumption should not be the case. We also remark and show how normalization is important in order to provide well-structured proofs.

We briefly suggest how to use the structural feature of sequent calculus in favor of producing explanations in natural language from proofs. As it was remarked at the introduction, the use of the cut-rule can provide shorter proofs. The cut-rule does not increase the complexity of the explanation, since it simply may provide more structure to the original proof. With the help of the results reported in this book one has a solid basis to build mechanisms to provide shorter and good explanation for $\mathcal{ALC}$ subsumption in the context of a KB authoring enviroment. The inclusion of the cut-rule, however, at the implementation level, is a hard one. Presently, there are approaches to include analytical cuts in Tableaux, as far as we know there is no research on how to extend this to $\mathcal{ALC}$ Tableaux. This puts our results in advantage when taking explanations, and the size of the proofs as well, into account. There are also other techniques, besides the use of the cut-rule, to produce short proofs in the sequent calculus, see [5, 7], that can be used in our context.

Althought not presented in this book, we have already proposed constructive (Intuitionistic) versions of ND$_{\mathcal{ALC}}$ and SC$_{\mathcal{ALC}}$ [8, 9]. This is a joint work with Edward Hermann, Valeria de Paiva and Mario Benevides. We extended the results obtained in this book proposing constructive versions of our systems and an aplication of this constructive version in the context of the problem of formalizing legal knowledge. In order to illustrate our approach, we formalized "Conflict of Laws in Space". This conflict happens when several laws can be applied, with different outcomes, to a case/situation depending on the place where the case occurs. Typical examples are those ruling the rights of a citizen abroad. The work with constructive DL was inspired by proposed constructive semantics for $\mathcal{ALC}$ in the literature [4, 13, 14].

9.2 Future Work

Future investigation must include the following topics:

- The extension of the calculi in order to deal with stronger Description Logics, mainly, $\mathcal{SHIQ}$ [1];
- The development of methods for proof explanation extraction from proofs;
- A proof of completeness for ND$_{\mathcal{ALCQI}}$ and SC$_{\mathcal{ALCQI}}$ should be obtained by extending the completeness proof for SC$_{\mathcal{ALC}}$;

References

1. Baader, F.: The Description Logic Handbook: Theory, Implementation, and Applications. Cambridge University Press, Cambridge (2003)
2. Baader, F., Peñaloza, R., Suntisrivaraporn, B.: Pinpointing in the description logic el $+ \mathcal{EL}^+$. In: Proceedings of the 30th annual German Conference on Advances in Artificial Intelligence, Osnabrück, Germany KI'07, Lecture Notes in Computer Science. vol. 4667/2007, pp. 52–67. Springer-Verlag, Berlin, Heidelberg (2007). ISBN:978-3-540-74564-8, doi:10.1007/978-3-540-74565-5_7

3. Bail, S., Horridge, M., Parsia, B., Sattler, U.: The justificatory structure of the ncbo bioportal ontologies. The Semantic Web-ISWC 2011, pp. 67–82 (2011)
4. Bozzato, L., Ferrari, M., Fiorentini, C., Fiorino, G.: A constructive semantics for ALC. In: Workshop on Description Logics, pp. 219–226. (2007)
5. Finger, M.: DAG sequent proofs with a substitution rule. In: Artemov, S., Barringer, H., d'Avila Garcez, A., Lamb, L., Woods, J. (eds.) We will show Them–Essays in honour of Dov Gabbay 60th birthday, vol. 1, pp. 671–686. Kings College Publications, Kings College, London (2005)
6. Gabbay, D.M.: Labelled deductive systems, vol. 1. Oxford University Press, Oxford (1996)
7. Gordeev, L., Haeusler, E., Costa, V.: Proof compressions with circuit-structured substitutions. In: Zapiski Nauchnyh Seminarov POMI, (2008) to appear
8. Haeusler, E.H., de Paiva, V., Rademaker, A.: Intuitionistic description logic and legal reasoning. In: Proceedings of International Workshop Data, Logic and Inconsistency with DEXA, (2011). (2011)
9. Haeusler, E.H., de Paiva, V., Rademaker, A.: Using intuitionistic logic as a basis for legal ontologies. Informatica e Diritto (Journal on Informatics and Law), 1–2:289–298 (2010)
10. Horridge, M., Parsia, B., Sattler, U.: Laconic and precise justifications in owl. The Semantic Web-ISWC 2008 pp. 323–338. (2008)
11. Ji, Q., Qi, G., Haase, P.: A relevance-directed algorithm for finding justifications of dl entailments. The Semantic Web pp. 306–320. (2009)
12. Kalyanpur, A., Parsia, B., Horridge, M., Sirin, E.: Finding all justifications of owl dl entailments. The Semantic Web pp. 267–280. (2007)
13. Mendler, M., Scheele, S.: Towards constructive DL for abstraction and refinement. In: F. Baader, C. Lutz, B. Motik (eds.) Proceedings of the 21st International Workshop on Description Logics, CEUR Workshop Proceedings, vol. 353, pp. 13–16. CEUR-WS.org, Dresden, Germany (2008). http://www.ceur-ws.org/Vol-353/MendlerScheele.pdf
14. de Paiva, V.: Constructive description logics: what, why and how. In: Context Representation and Reasoning. Riva del Garda (2006)